"十四五"职业教育国家规划教材

数控技术应用专业英语

新世纪高职高专教材编审委员会 组编
主　编 李桂云 冯艳宏
副主编 王春武 雷　洁

第五版

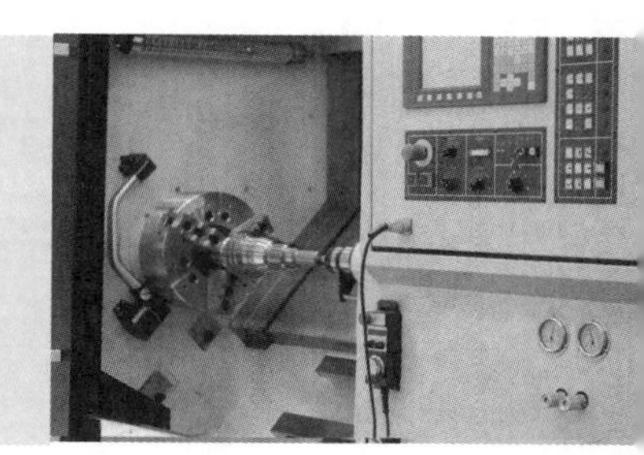

大连理工大学出版社

图书在版编目(CIP)数据

数控技术应用专业英语 / 李桂云，冯艳宏主编. -- 5版. -- 大连：大连理工大学出版社，2022.1(2023.12重印)
ISBN 978-7-5685-3700-1

Ⅰ.①数… Ⅱ.①李… ②冯… Ⅲ.①数控技术－英语－教材 Ⅳ.①TP273

中国版本图书馆CIP数据核字(2022)第021894号

大连理工大学出版社出版
地址:大连市软件园路80号　邮政编码:116023
电话:0411-84708842　邮购:0411-84708943　传真:0411-84701466
E-mail:dutp@dutp.cn　URL:https://www.dutp.cn
大连图腾彩色印刷有限公司印刷　　大连理工大学出版社发行

幅面尺寸:185mm×260mm　印张:11.5　字数:277千字
2008年4月第1版　2022年1月第5版
2023年12月第4次印刷

责任编辑:陈星源　责任校对:刘　芸
封面设计:张　莹

ISBN 978-7-5685-3700-1　定　价:45.00元

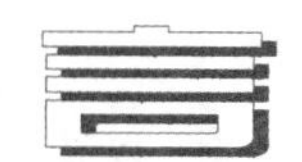

前言

《数控技术应用专业英语》(第五版)是“十四五”职业教育国家规划教材、“十三五”职业教育国家规划教材及“十二五”职业教育国家规划教材。

本教材顺应高等职业教育课程改革的发展要求,落实立德树人根本任务,发挥专业课程育人作用。教材根据职业院校数控技术相关专业教学标准,经过广泛的企业调研和对数控技术专业毕业生专业英语能力要求的充分论证,在深入了解各职业院校师生教学需求的基础上精心编写而成。

本教材从高职院校专业英语课时不同的需求和分层教学需求出发,将教材分为专业精读、专业泛读和职场应用三大部分,并有机融入课程思政。专业精读按照数控技术专业职业岗位群创设数控车削加工、数控加工中心加工和数控机床维护与故障诊断技术三个模块,专业泛读包括数控加工技术基础和先进制造技术两个模块,职场应用包括英文说明书阅读、求职申请撰写和展会对话等职场实际应用内容。

本次教材修订突出以下特色:

1. 工匠精神,立德树人,铸魂育人

本次修订全面贯彻党的二十大精神,增加了与教学内容相对应的素养提升案例,涵盖了执着专注、精益求精等工匠精神的内容,将思政元素渗透、融入教材中,起到潜移默化的育人效果,为青年走好人生之路提供强大的精神指引,对促进新时代青年成长成才、引导青年为实现中华民族伟大复兴不懈奋斗具有重要的时代意义。

2. 多措并举,提升听说读写能力

新版教材增加对话音频、课文音频和拓展阅读音频二维码。以对话导入新任务,对话音频涵盖正文核心知识点词汇,扫描二维码可以听对话,锻炼学生的听、说能力。课文音频和拓展阅读音频可以锻炼学生的阅读能力与听力,结合核心词汇进行写作练习。教材设计可以实现综合提升学生的听、说、读、写四项技能的全面训练。

3. 精细设计,适应不同教学需求

随着教育教学改革的不断深入,各地区、各院校对专业英语课程教学课时的分配相差较大,为了更好地适应不同院校的需求，体现分层教学的需要，本教材设计了6个模块，既优化了知识重叠问题，又对教材中涉及的基础知识进行系统化设计，还有利于实施分层教学。

4.资源丰富，满足师生多样化需求

为了方便教师教学和学生自学，本教材配有对话音频、课文音频、拓展阅读音频和扫码阅读，以及课程标准、课程设计方案、教案、课件、习题参考答案及模拟试卷等形式多样、内容丰富的教学资源。

本教材可供高等职业院校机械制造及自动化、机械设计与制造、机电一体化技术、数控技术、模具设计与制造等专业教学使用，也可以作为加工制造业的技术人员学习专业英语的参考书。

本教材由天津工业职业学院李桂云、冯艳宏任主编，天津电气科学研究院有限公司教授级高工王春武、天津机电职业技术学院雷洁任副主编，天津工业职业学院王晓霞和李焱、秦皇岛职业技术学院王英刚、兰州职业技术学院王新陇任参编。具体编写分工如下：李桂云编写模块1的任务2和3、模块2的任务1、模块3的任务1、模块6的任务2和3；冯艳宏编写模块2的任务2和3、模块3的任务2、模块4的任务3和5、模块5的任务4、模块6的任务1以及专业精读对话和课程思政内容；王春武编写模块4的任务2的中文；雷洁编写模块5的任务1；王晓霞编写模块1的任务1和模块4的任务4；李焱编写模块5的任务3和模块4的任务1的中文；王英刚编写模块4的任务1和任务2的英文，并对全书英文表达进行润色；王新陇编写模块5的任务2和模块2的任务2的部分内容。全书由李桂云和冯艳宏负责统稿和定稿。

在编写本教材的过程中，我们参考、引用和改编了国内外出版物中的相关资料以及网络资源，在此对这些资料的作者表示诚挚的谢意。请相关著作权人看到本教材后与出版社联系，出版社将按照相关法律的规定支付稿酬。

尽管我们在教材特色的建设方面做出了许多努力，但由于编者水平有限，教材中仍可能存在一些疏漏和不妥之处，恳请各教学单位和读者在使用本教材时多提宝贵意见，以便下次修订时改进。

编　者

所有意见和建议请发往：dutpgz@163.com
欢迎访问职教数字化服务平台：https://www.dutp.cn/sve/
联系电话：0411-84707424　84708979

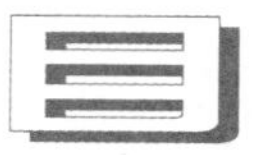

目录

素养提升

序号	名称	二维码	对应任务	序号	名称	二维码	对应任务
1	练就“金手指”，铸就大工匠		1-2	9	传承钱学森精神		4-2
2	6S管理		1-3	10	坚守初心，砥砺前行		4-3
3	打造中国高端制造的“金刚钻”		2-1	11	从倪志福钻头到群钻		4-4
4	匠心传承，技能报国		2-2	12	手眼配合，精准测量		4-5
5	工匠绝技，精益求精		2-3	13	中国机器人之父		5-2
6	打造“蛟龙”的大国工匠		3-1	14	“绿色工厂”引领绿色制造新高度		5-4
7	黄维祥：数控能手的“三多”秘诀		3-2	15	用极致书写精密人生		6-3
8	中国高温合金之父		4-1				

Module 1
CNC Lathe Machining

Task 1 Basic Knowledge of CNC Lathe

Speaking Activity

A: Excuse me, may I ask you a question, Mr. Li?

B: Of course.

A: I often confuse about the kinds of CNC lathe. Can you explain them to me?

B: OK, let me tell you. CNC lathes are classified as horizontal and vertical CNC lathes. Horizontal CNC lathes are for a long axial dimension or a small disc-type workpiece turning. Vertical CNC lathes are used to turn large diameter disc-type workpieces.

A: Do you know about the mechanical components of a CNC lathe?

B: I know some of them. They are bed, headstock, three-jaw chuck, turret, tail-stock, operation panel, protective shield and guide way, etc..

A: What's this?

B: It's a three-jaw chuck and a workpiece can be mounted on it.

A: What's that?

B: It's a bed. These are the main structures of the lathe.

A: Oh, I see. Thanks.

B: It is my pleasure.

Part A Technical Reading

1. General Introduction about a CNC Lathe

The CNC concept was proposed in the late 1940s by John Parsons. A

CNC lathe is a computer numerical control lathe. It is primarily used to rotate round workpieces with turning, boring, drilling, reaming, tapping and other machining processes. Firstly, the information required for machining workpiece is recorded by the program that is stored in some media when the CNC lathe is working, and then the program is input to the numerical control device. Processed by the CNC device, commands and control signals are issued to servo system to drive the machine, coordinate the machine movements and make it produce a series of machine movements such as the main motion and feed motion, to complete workpiece machining. When changing the shape of the workpiece, we just modify the CNC program to continue the machining.

CNC lathes are classified as horizontal (Fig. 1-1-1) and vertical (Fig. 1-1-2) CNC lathes. Horizontal CNC lathes are for a long axial dimension or a small disc-type workpiece turning. Vertical CNC lathes are used to turn a large diameter disc-type workpiece turning.

Fig. 1-1-1 Horizontal CNC Lathe

Fig. 1-1-2 Vertical CNC Lathe

CNC lathes have the following main features:

(1) Short transmission chain and high precision machining;

(2) Simple headstock, simplified structure and greatly increased stiffness;

(3) Sufficient cooling and more rigorous protection;

(4) High production efficiency;

(5) Reducing labor intensity and improving working conditions.

2. Components of a CNC Lathe

The mechanical components of a CNC lathe are bed, headstock, three-jaw chuck, turret, tailstock, operation panel, protective shield and guide way, etc. (Fig. 1-1-3).

CKA6150 lathe has such configurations:

(1) The bed, headstock and saddle plinth, etc. are molded with resin sand casting;

(2) Feeding system uses servo motors, precise ball screws and high rigidity compound bearings that ensure accurate positioning and efficient driving;

(3) High spindle speed with wide variable speed ranges, low noise;

(4) Flexible control panel enables convenient tool setting;

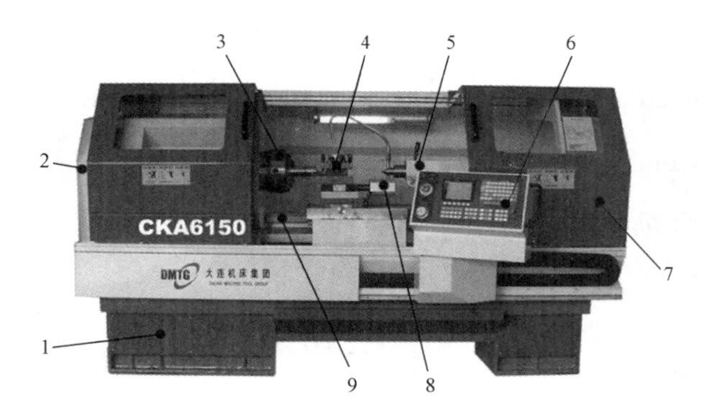

Fig. 1-1-3 Mechanical Components of a CNC Lathe

1—bed; 2—headstock; 3—three-jaw chuck; 4—turret; 5—tailstock;
6—operation panel; 7—protective shield; 8—turret motor; 9—guide way

(5) Vertical 4-position turret, horizontal 6-position turret and quick change tool post, etc. are available for your selection;

(6) FANUC, SIEMENS, FAGOR, etc. control systems can be appointed.

New Words and Phrases

lathe /leɪð/	*n*. 车床
bore /bɔː/	*vt*. 使(孔)变大,扩宽(孔)的直径
drill /drɪl/	*vt*. 钻(孔),打(眼)
ream /riːm/	*vt*. 铰孔,扩孔
tap /tæp/	*vt*. 攻丝
servo system	伺服系统
horizontal /ˌhɒrɪˈzɒntəl/	*adj*. 水平的,与地平线平行的,横的
vertical /ˈvɜːtɪkəl/	*adj*. 垂直的,竖的
spindle /ˈspɪndl/	*n*. 主轴
headstock /ˈhedstɒk/	*n*. 主轴箱
guide way	导轨
turret /ˈtʌrət/	*n*. 刀架
tailstock /ˈteɪlstɒk/	*n*. 尾架,尾座,顶针座
cast /kɑːst/	*vi*. 铸造,铸成
resin /ˈrezɪn/	*n*. 树脂,合成树脂
plinth /plɪnθ/	*n*. 底座,基座

ball screw	滚珠丝杠
hydraulic /haɪˈdrɒlɪk/	*adj*. 液力的，液压的
program /ˈprəʊgæm/	*n*. 程序
workpiece /ˈwɜːkˌpiːs/	*n*. 工件

Part B Practice Activity

Ⅰ. Fill in the blanks according to the text.

1. The CNC concept was proposed in the ________ 1940s.

2. It is primarily used to rotate round workpieces with ________, boring, ________, reaming, tapping and other machining processes.

3. Machine movements of CNC lathe include the ________ motion and ________ motion.

4. Vertical CNC lathes are used to turn ________ workpieces.

5. The main components of a CNC lathe are ________, ________ and ________, etc..

Ⅱ. Match A with B.

A	B
horizontal CNC lathe	立式数控车床
workpiece	程序
vertical CNC lathe	生产率高
high production efficiency	卧式数控车床
program	工件

Ⅲ. Choose the best answers.

1. CNC lathe is used to rotate ( ) workpiece.

A. heart B. triangle C. diamond D. round

2. CNC lathes are classified as horizontal and ( ) CNC lathe.

A. slanting B. vertical C. parallel D. transverse

3. Headstock box of CNC lathe is very ( ).

A. big B. small C. simple D. complex

4. Transmission chain of CNC lathe is ( ).

A. short B. long C. big D. brief

5. Which is not the CNC lathe's main feature? ( )

A. High production efficiency. B. Simple spindle.

C. Improving labor intensity. D. Cooling fully.

Ⅳ. Answer the following questions briefly according to the text.

1. What do we need to do to make the CNC lathe continue machining when changing the shape of the workpieces?

____

2. What type of workpieces can be turned by horizontal CNC lathe?

____

3. What materials are used to mold headstock and saddle plinth?

____

Professional Situation Simulation

Please write a composition in about 150 words according to the following key words.

Key Words: *headstock*, *three-jaw chuck*, *turret*, *tailstock*, *operation panel*, *guide way*

Work Sheet

....

....

....

....

....

....

....

....

....

....

Part C Broaden Your Horizon

The use of CNC machines will still increase in the future. Not only in industrial production but also in small workshops, conventional machines will be replaced by CNC machines. The application of CNC technics is not bound to the classic machines such as lathes, milling machines or to the metal working area. One could say, nearly every day a new application of CNC technics is realized. Practically all occupations such as technical designer, technical manager or salesman, skilled worker, method engineer and controller, etc. will be confronted with CNC technology in many ways.

模块 1 数控车削加工

任务 1 认识数控车床

1. 数控车床概述

数控概念是约翰·帕森斯于 20 世纪 40 年代后期提出的。数控车床是计算机数字控制车床,主要用于旋转体零件的车削、镗削、钻削、铰削、攻丝等加工。数控车床工作时,首先要把加工工件需要的信息以程序的形式记录下来,存储在载体上,然后将程序输入到数控装置中。由数控装置处理程序,发出指令和控制信号给伺服系统以驱动机床、协调机床动作,使其产生主运动和进给运动等一系列机床运动,完成工件的加工。当改变工件的形状时,只需修改数控程序,就可以继续加工。

数控车床分为卧式数控车床(图 1-1-1)和立式数控车床(图 1-1-2)。卧式数控车床用于轴向尺寸较长或小型盘类零件的车削加工。立式数控车床用于回转直径较大的盘类零件的车削加工。

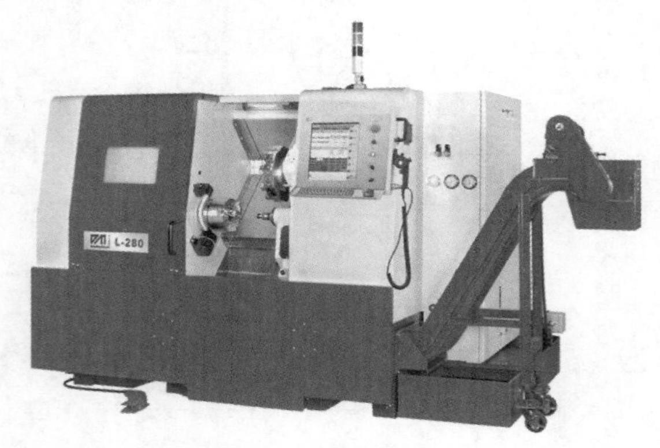

图 1-1-1 卧式数控车床

图 1-1-2 立式数控车床

数控车床的主要特点如下:

(1)传动链短,加工精度高;

(2)主轴箱简单,结构简化,刚度显著提高;

(3)冷却充分,防护更严密;

(4)生产率高;

(5)减轻劳动强度,改善劳动条件。

2. 数控车床的组成部件

数控车床的机械部件有床身、主轴箱、三爪卡盘、刀架、尾座、操作面板、防护罩和导轨等(图 1-1-3)。

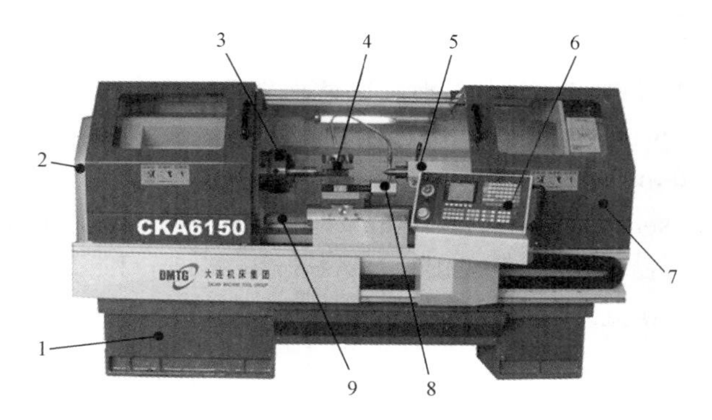

图 1-1-3 数控车床的机械部件

1—床身；2—主轴箱；3—三爪卡盘；4—刀架；5—尾座；6—操作面板；7—防护罩；8—刀架电动机；9—导轨

CKA6150 型车床有以下配置：

(1)床身、主轴箱、床鞍等基础件均采用树脂砂型铸造；

(2)进给系统采用伺服电动机、精密滚珠丝杠、高刚性精密复合轴承结构，以保证定位准确、传动效率高；

(3)主轴转速高，调速范围宽，整机噪声低；

(4)柔性控制面板便于操作者对刀；

(5)刀架有立式四工位、卧式六工位、快换刀架等多种形式可供选择；

(6)控制系统有 FANUC、SIEMENS、FAGOR 等品牌供选择。

Task 2 CNC Program for Lathe

Speaking Activity

A: Hi, how are you doing?

B: Not so good.

A: What's wrong with you? Can I help you?

B: Yes, I don't know the difference between the machine coordinate system and the workpiece coordinate system.

A: The machine coordinate system is the inherent coordinate system of CNC machine and has a fixed coordinate origin, while the workpiece coordinate system is the coordinate system used in programming, also known as the programming coordinate system.

B: Who sets the machine zero point for each machine?

A: The manufacturer.

B: I am not very clear about the program zero. Can you explain it to me?

A:Of course. The origin point for each axis is commonly called the program zero. The placement of program zero is determined by the programmer.

B:Can it be placed anywhere?

A:Yes. The wise selection of a program zero point will make programming much easier. The program zero of a lathe is generally set in the midpoint of the workpiece's right surface.

B:Thanks.

A:It is my pleasure.

Part A Technical Reading

1. Preparing for CNC Lathe Program

(1)Coordinate Systems

● Machine Coordinate System(MCS)

The machine manufacturer sets a machine zero point for each machine. The machine zero point is the origin of the machine coordinate system.

● Workpiece Coordinate System(WCS)

When the operator wants to determine the position of the workpiece on the machine, he must set the workpiece coordinate systems. Several workpiece coordinate systems can be set for one workpiece.

● Local Coordinate System(LCS)

If the dimension of a workpiece is too big, the user can set another coordinate system in a local area of the workpiece, which is a local coordinate system.

The three coordinate systems are shown in Fig. 1-2-1.

(2)Program Zero Point

The origin point for each axis is commonly called the program zero point, also called workpiece zero, workpiece origin or zero point. One of the main benefits of using the rectangular coordinate system is that many of the coordinates used within the program can be taken right from the drawings.

The placement of the program zero point is determined by the programmer. The program zero point could be placed anywhere. The wise selection of the program zero point will make programming much easier. You should always make your program zero point a location on the datum surfaces of your workpiece. There is a program zero point for each axis.

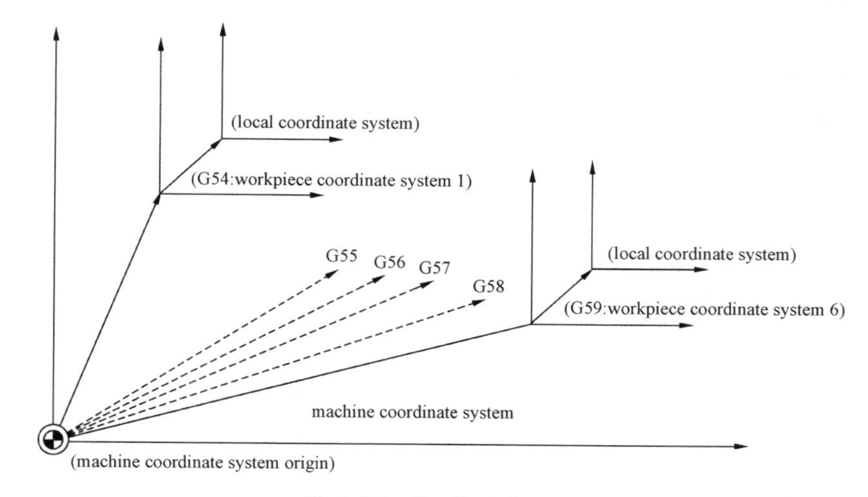

Fig. 1-2-1 Coordinate Systems

The program zero point of a lathe is generally set in the midpoint of the workpiece's right surface(Fig. 1-2-2).

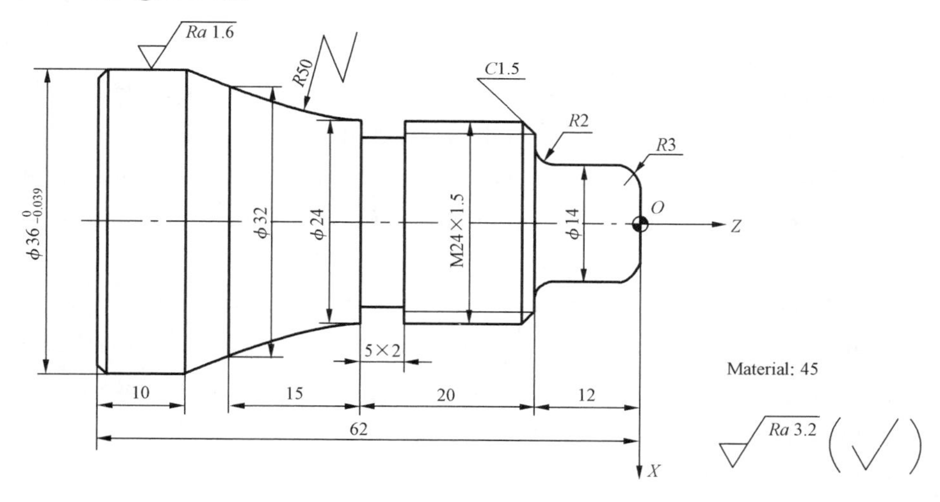

Fig. 1-2-2 Program Zero Point of a Lathe Workpiece

(3) Absolute and Incremental Mode

When you are specifying coordinates from the program zero point, it is called the absolute mode of programming. For machining center(CNC lathe) programming, the absolute mode is specified by a G90(X, Z) word in the program. In the incremental system, every movement refers to the previously dimensioned position. If the programmer makes a mistake in incremental program, every incremental movement from that point will be incorrect. If the programmer makes the same mistake in an absolute program, only one movement will be incorrect. Any motion can be commanded in either the absolute or the incremental mode.

(4) Program Configuration

Generally speaking, there are two types of programs: the main program and the subprogram. The main program contains a series of commands for machining workpieces. The subprogram can be called by the main program or another subprogram.

Either the main program or the subprogram contains three parts: the program number, the program content and the program end(Fig. 1-2-3).

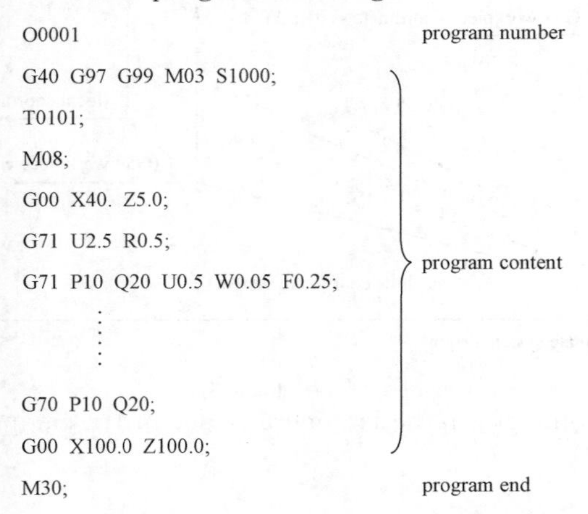

Fig. 1-2-3 Program Configuration

The program number usually starts with "O"(or "%"). The number after "O"(or "%") is the program number. Program numbers range from 0001 to 9999.

The content of the program is the core of the whole program. It is made up of many blocks. These blocks control the movements that the CNC machine is to execute.

The M02 or M30 command is used to stop the whole program.

A CNC program consists of one or more blocks of commands. A block is a complete line of information to the CNC machine. It consists of one word or an arrangement of words. Blocks may vary in length. Thus, a programmer needs to master all the words required to execute a particular machine function.

(5) The F, S, M and T Codes

The F code controls the feed rate(Fig. 1-2-4). It can be expressed as feed per minute or feed per revolution. An F code is a modal code and remains in effect in a program for all subsequent tool movements. The feed rate can be changed frequently in a program, as needed.

The S code controls the spindle speed(Fig. 1-2-5). The selected speed value right follows the S address. The S command should be given together with the spindle revolution command(M03 or M04). For example, "M03 S1600;" directs the control to set the CNC machine spindle 1 600 r/min.

The M codes are called the miscellaneous words and are used to control miscellaneous functions of the machine. Such functions include turning the spindle on/off, starting/stopping the machine, turning on/off the coolant, changing the tool and returning to the start of the program.

The T code is used to specify the tool number(Fig. 1-2-6).

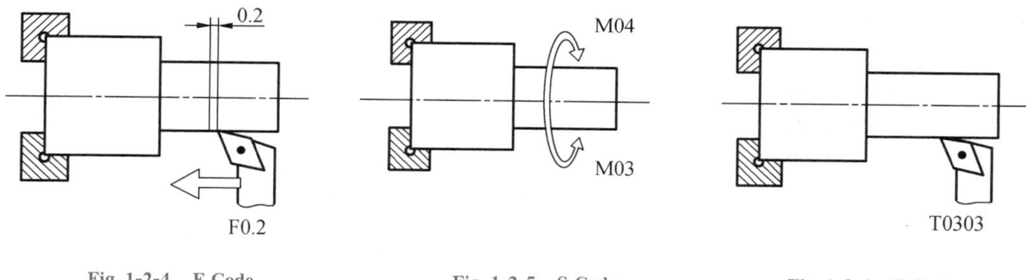

Fig. 1-2-4 F Code

Fig. 1-2-5 S Code

Fig. 1-2-6 T Code

(6) The G Codes

The G codes are also called preparatory codes or words. A preparatory code is designated by the address G followed by digits to specify the mode in which a CNC machine moves along its programmed axes. G codes are usually placed at the beginning of a block. G codes are usually classified into two types. The modal G code will remain in effect for all subsequent blocks unless replaced by another G code of the same group. The non-modal G codes will only affect the block in which it appears.

2. Program Codes

(1) Basic Program Codes

- G00(Rapid Traverse/Positioning)

The G00 command moves a tool to the position in the workpiece system specified with an absolute or an incremental command at a rapid traverse rate. In the absolute command, the coordinate value of the end point is programmed. In the incremental command, the distance of the tool move is programmed. The rapid traverse rate cannot be specified in the address F. The format of command is "G00 X(U)__ Z(W)__;". The coordinate of the end point is specified by X(U) and Z(W). X or Z is expressed as an absolute coordinate value. U or W is expressed as an incremental coordinate value.

- G01(Linear Interpolation)

The G01 command moves a tool along a line to the specified position at the feed rate specified in F code. The feed rate specified in F code is effective until a new value is specified. It doesn't need to be specified for each block. G01 and G00 are modal G codes. The format of command is "G01 X(U)__ Z(W)__ F__;". F is expressed as rate of tool feed.

- G02/G03(Circular Interpolation)

The command will move a tool along a circular arc. The format of command is "G02(G03) X(U)__ Z(W)__ R__(or I__ K__) F__;". Clockwise(G02) and counterclockwise(G03) on the *XZ* plane are defined when the *XZ* plane is viewed in the positive to negative direction of the *Y* axis in the Cartesian coordinate system, as shown in Fig. 1-2-7.

The end point of an arc is specified by X(U)and Z(W). For the incremental value, the distance of the end point that is viewed from the start point of the arc is specified.

The arc center is specified by addresses I and K(for the coordinates of the *X* and *Z* axes, respectively). The numerical value following I and K, however, is a vector component in which the arc center is seen from the start point, as shown in Fig. 1-2-8.

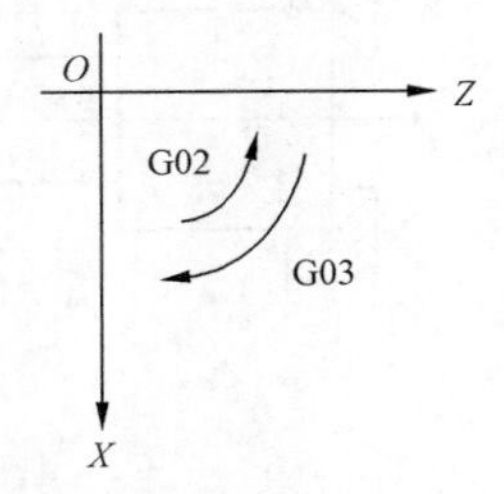

Fig. 1-2-7 Circular Interpolation

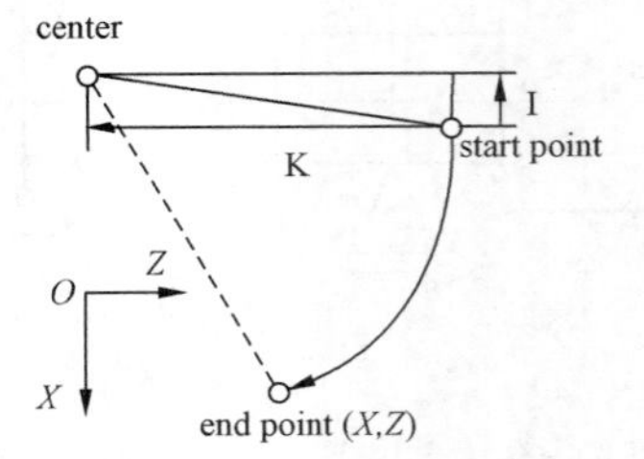

Fig. 1-2-8 I and K

● G04(Dwell)

The execution of the next block is delayed by the specified time. The format of command is "G04 X __;" or "G04 U __;" or "G04 P __;". X/U/P is expressed as specifying a time (decimal point permitted/decimal point permitted/decimal point not permitted).

(2) Tool Compensation Codes

Most controls use three G codes for tool nose radius compensation. G41 is used to instate a tool nose radius compensation left. G42 is used to instate a tool nose radius compensation right. G40 is used to cancel tool nose radius compensation.

How to use G41 and G42 for turning operation is shown in Fig. 1-2-9. Once tool nose radius compensation is properly instated, the cutter will be kept on the left side or right side of all surfaces until the G40 code cancels tool nose radius compensation.

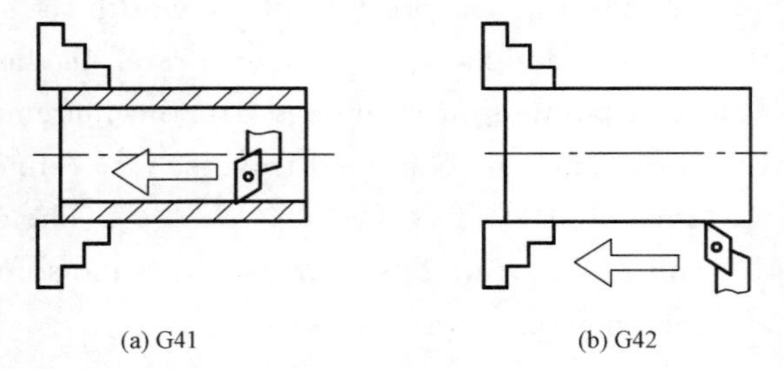

(a) G41 (b) G42

Fig. 1-2-9 G41 and G42

(3) Cycle Codes

● G71(Stock Removal in Turning)

This canned cycle will rough out material on a part given the finished part shape. All that a programmer needs to do is define the shape of a part by programming the finished tool path and then submitting the path definition to G71 by means of a PQ block designation. The following is the format of program code:

G71 U(Δd) R(e);

G71 P(ns) Q(nf) U(Δu) W(Δw) F(f) S(s) T(t);

In that Δd—depth of cut(radius designation);

e—escaping amount;

ns—sequence number of the first block for the program of the finishing;

nf—sequence number of the last block for the program of the finishing;

Δu—distance and direction of finishing allowance in *X* direction(diameter designation);

Δw—distance and direction of finishing allowance in *Z* direction;

f, *s*, *t*—any F, S or T function contained in blocks *ns* to *nf* in the cycle is ignored, and the F, S, or T function in this G71 block is effective.

◎ G73(Pattern Repeating in Turning)

By this cutting cycle, it is possible to efficiently cut workpiece whose rough shape has already been made by a rough machining, forging or casting method, etc.. The following is the format of program code:

G73 U(Δi) W(Δk) R(*d*);

G73 P(*ns*) Q(*nf*) U(Δu) W(Δw) F(*f*) S(*s*) T(*t*);

In that Δi—distance and direction of relief in *X* direction(radius designation);

Δk—distance and direction of relief in *Z* direction;

d—the number of division.

Other meanings are similar to that in G71.

◎ G70(Finishing Cycle in Turning)

After rough cutting by G71 or G73, the G70 finishing cycle can be used for finishing cutting. The following is the format of program code:

G70 P(*ns*) Q(*nf*);

In that, the meanings of *ns* and *nf* are similar to G71 and G73.

3. Application of CNC Program

Part drawing is shown in Fig. 1-2-10. Part program and explanation are shown in Chart 1-2-1.

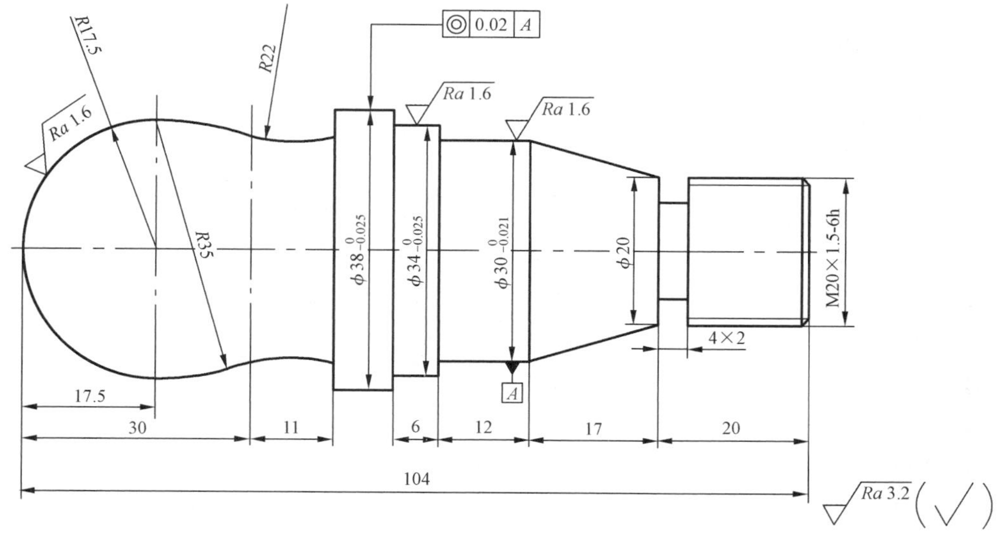

Fig. 1-2-10 Part Drawing

Chart 1-2-1 Part Program and Explanation

Program	Explanation
O0001	No. 0001 program
G40 G97 G99 M03 S600 F0.2;	cancel tool nose radius compensation, set constant spindle speed control, spindle CW at 600 r/min, feed rate is 0.2 mm/r
T0101;	call No. 1 tool and its offset
G00 G42 X40.0 Z5.0;	rapid positioning to point(X40.0, Z5.0), call tool nose radius compensation
G71 U1.5 R0.5;	call stock removal in turning
G71 P10 Q20 U0.5 W0.05;	
N10 G00 X0.0;	finishing shape
G01 Z0.0;	
X19.85 C1.5;	
Z−20.0;	
X20.0;	
X30.0 W−17.0;	
W−12.0;	
X34.0;	
W−6.0;	
X38.0;	
W−10.0;	
N20 X40.0;	
G00 G40 X100.0 Z100.0;	rapid positioning to point(X100.0, Z100.0), cancel tool nose radius compensation
M03 S1000 T0101 F0.1;	spindle CW at 1 000 r/min, call No. 1 tool and its offset, feed rate is 0.1 mm/r
G00 G42 X40.0 Z5.0;	rapid positioning to point(X40.0, Z5.0), call tool nose radius compensation
G70 P10 Q20;	call finishing cycle in turning
G00 G40 X100.0 Z100.0;	rapid positioning to point(X100.0, Z100.0), cancel tool nose radius compensation
M03 S400 T0202 F0.08;	spindle CW at 400 r/min, call No. 2 tool(grooving tool) and its offset, feed rate is 0.08 mm/r
G00 Z−20.0;	rapid positioning to point(X100.0, Z−20.0)
X22.0;	rapid positioning to point(X22.0, Z−20.0)
G01 X16.0;	linear interpolation to point(X16.0, Z−20.0)
G04 X2.0;	call dwell code, dwell time is 2 seconds
X22.0;	linear interpolation to point(X22.0, Z−20.0)
G00 X100.0 Z100.0;	rapid positioning to point(X100.0, Z100.0)
M03 S400 T0303;	spindle CW at 400 r/min, call No. 3 tool(thread tool) and its offset
G00 X21.0 Z5.0;	rapid positioning to point(X21.0, Z5.0)

continued

Program	Explanation
G76 P020060 Q100 R0.1;	call the multiple thread cutting cycle
G76 X18.05 Z−18.0 P975 Q400 F1.5;	
G00 X100.0 Z100.0;	rapid positioning to point(*X*100.0,*Z*100.0)
M30;	end of program and reset
O0002	No.0002 program
G40 G97 G99 M03 S600 F0.2;	cancel tool nose radius compensation, set constant spindle speed control, spindle CW at 600 r/min, feed rate is 0.2 mm/r
T0101;	call No.1 tool and its offset
G00 G42 X40.0 Z5.0;	rapid positioning to point(*X*40.0,*Z*5.0), call tool nose radius compensation
G73 U20.0 W0 R10.0;	call pattern repeating in turning
G73 P10 Q20 U0.5 W0.05;	
N10 G00 X0.0;	finishing shape
G01 Z0.0;	
G03 X35.0 Z−17.5 R17.5;	
G03 X30.567 Z−30.0 R17.5;	
G02 X30.567 W−11.0 R22.0;	
N20 X40.0;	
G00 G40 X100.0 Z100.0;	rapid positioning to point(*X*100.0,*Z*100.0), cancel tool nose radius compensation
M03 S1000 T0101 F0.1;	spindle CW at 1 000 r/min, call No.1 tool and its offset, feed rate is 0.1 mm/r
G00 G42 X40.0 Z5.0;	rapid positioning to point(*X*40.0,*Z*5.0), call tool nose radius compensation
G70 P10 Q20;	call finishing cycle in turning
G00 G40 X100.0 Z100.0;	rapid positioning to point(*X*100.0,*Z*100.0), cancel tool nose radius compensation
M30;	end of program and reset

New Words and Phrases

coordinate system	坐标系
reference point	参考点
be defined as	被认定为……,被定义为……
machine zero point	机床零点
manufacturer /ˌmænjuˈfæktʃərə/	*n.* 厂家,制造商
origin /ˈɒrɪdʒɪn/	*n.* 起点,原点
determine /dɪˈtɜːmɪn/	*v.* 确定,决定

local area	局部地区
axis /ˈæksɪs/	*n.* 轴(复数 axes)
program zero point	程序零点
datum /ˈdeɪtəm/	*n.* 基准,基准线,基准面
absolute /ˈæbsəluːt/	*adj.* 完全的,绝对的
incremental /ˌɪnkrɪˈmentəl/	*adj.* 增加的,增量的
program configuration	程序结构
main program	主程序
subprogram /sʌbˈprəʊɡræm/	*n.* 子程序
program content	程序内容
core /kɔː/	*n.* 核心
preparatory function	准备功能
modal /ˈməʊdl/	*adj.* 形式的,样式的,模态的
subsequent /ˈsʌbsɪkwənt/	*adj.* 后来的,其次的
express /ɪkˈspres/	*v.* 表示,表达
feed per minute	每分钟进给(量)
miscellaneous /ˌmɪsəˈleɪniəs/	*adj.* 辅助的,混杂的
program code	编程代码
linear /ˈlɪniə/	*adj.* 线性的
interpolation /ɪnˌtɜːpəˈleɪʃn/	*n.* 插补
counterclockwise /ˌkaʊntəˈklɒkwaɪz/	*adj.* 逆时针方向的
Cartesian coordinate system	笛卡儿坐标系
vector /ˈvektə/	*n.* 矢量
dwell /dwel/	*v.* 停歇
compensation /ˌkɒmpenˈseɪʃn/	*n.* 补偿
canned cycle code	固定循环代码
stock /stɒk/	*n.* 原料,材料
pattern /ˈpætn/	*n.* 样式
rough machining	粗加工
thread /θred/	*n.* 螺纹
constant spindle speed control	主轴恒转速控制

Part B Practice Activity

Ⅰ. Match A with B.

A	B
圆弧插补	coordinate system
粗加工	tool nose radius compensation
直线插补	main program
参考点	reference point
退刀量	circular interpolation
主程序	escaping amount
坐标系	rough machining
刀尖半径补偿	linear interpolation

Ⅱ. Answer the following questions briefly according to the text.

1. What does the program number usually start with?

____

2. Who determines the program zero point?

____

3. What command is used to end the program?

____

4. What's the disadvantage of the incremental programming?

____

____

5. Who sets the machine zero point for each machine?

____

6. What are the G codes?

____

7. What's absolute mode of programming?

____

Ⅲ. Mark the following statements with T(true) or F(false).

(　　) 1. There are two types of program: the main program and the subprogram.

(　　) 2. Program blocks usually start with "P"(or "%").

(　　) 3. The content of the program is made up of many program blocks. Program blocks are made up of many words.

(　　) 4. G codes are usually classified into two main types.

(　　) 5. The S code controls the cutting feed.

(　　) 6. The machine coordinate system is set by the user.

(　　) 7. The logical selection of a program zero point will make programming much easier.

(　　) 8. In a workpiece coordinate system, in order to program easily a machine coordinate system can be set.

Ⅳ. Fill in the blanks according to the text.

1. M codes are used to control ____________ function of the machine.

2. The arc center is specified by addresses I and K for the ________________ axes, respectively.

3. In the absolute command, coordinate value of the end point is programmed. In the incremental command, the ________ tool move is programmed.

4. The feed rate specified in F code is effective until ________________________.

5. G71 command will ________________ material on a part given the finished part shape.

6. The three coordinate systems are ________________, ________________ and ________________ coordinate system.

7. The placement of ____________ is determined by the programmer.

Professional Situation Simulation

Please write a composition in about 150 words according to the following key words.

Key Words: *coordinate system*, *program code*, *program zero point*, *rough machining*

Work Sheet

Part C Broaden Your Horizon

On a workpiece-rotating machine, such as a lathe, the Z axis is parallel to the spindle. On a tool-rotating machine, such as a vertical milling or boring machine, the Z axis is perpendicular to the workpiece and the positive motion increases the distance between the tool and the workpiece.

On a lathe, the X axis is the direction of the workpiece's diameter. On a vertical milling machine, the X axis is parallel to the table. The positive X axis points to the right when the programmer is facing the machine.

Translation

任务 2 数控车削编程

1. 数控车削编程准备

(1)坐标系

◎机床坐标系(MCS)

机床制造商为每一台机床设置一个机床零点。机床零点是机床坐标系的原点。

◎工件坐标系(WCS)

当操作者想要确定零件在机床上的位置时,必须建立工件坐标系。一个工件可以设置几个工件坐标系。

◎局部坐标系(LCS)

如果工件的尺寸太大,则用户可以在工件上的某一局部区域设置另一个坐标系,这个坐标系就是局部坐标系。

三种坐标系如图 1-2-1 所示。

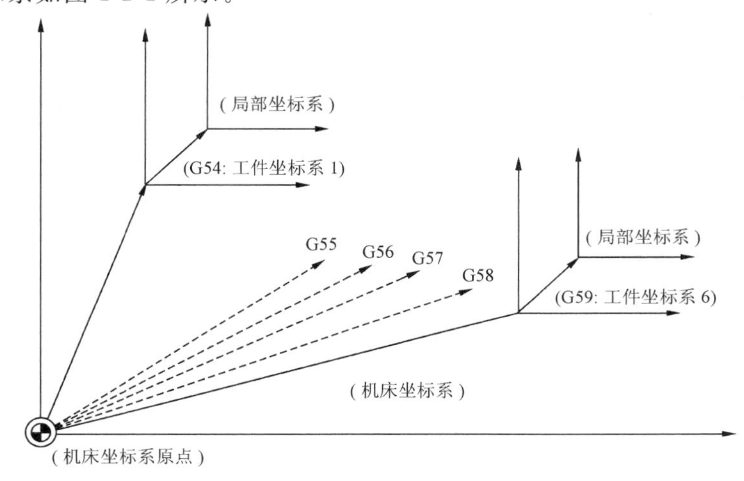

图 1-2-1 坐标系

(2)程序零点

每个坐标轴的原点通常称为程序零点,也称为工件零点、工件原点或零点。使用直角坐标系编程的一个主要好处是,程序中要用到的许多坐标都可以直接从图纸中得到。

程序零点的位置由编程者确定。程序零点可以被设定在任何位置,合理选择程序零点会使编程更加简便。程序零点应选在工件的尺寸基准上。每个坐标轴都有一个程序零点。

车床的程序零点一般设在工件右端面中点处(图 1-2-2)。

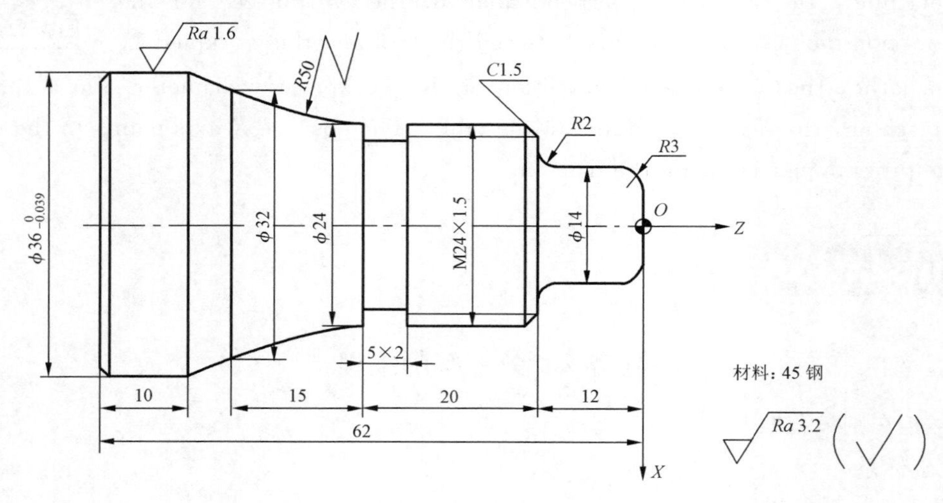

图 1-2-2　车床工件的程序零点

(3)绝对与增量模式

以程序零点确定坐标值称为绝对模式编程。对于加工中心(数控车床)编程,用 G90 (*X*,*Z*)表示绝对模式。在增量系统中,每个运动都是相对于前一个位置的。如果编程者在增量编程中把某点坐标值弄错,那么从那点后的每一个增量运动都是不正确的。如果编程者在绝对编程中出现同样的错误,则只有一个运动是不正确的。任何运动都可以用绝对或增量模式来表示。

(4)程序结构

一般来说,程序类型有两种:主程序和子程序。主程序包括一系列加工工件的指令,子程序可以被主程序或另外一个子程序调用。

不论是主程序还是子程序,都由三部分组成:程序号、程序内容和程序结束(图 1-2-3)。

程序号通常以“O”(或“%”)开始。“O”(或“%”)后面的数字是程序号。程序号的范围从 0001 至 9999。

程序内容是整个程序的核心,它由许多程序段组成,这些程序段控制数控机床要执行的运动。

M02 或 M30 指令用于停止整个程序。

一个数控程序包含一个或多个程序段。一个程序段是数控机床信息的一个完整行,它包含一个字或一系列字。程序段的长度可以变化,因此,编程者需掌握执行特定的机床功能所需的所有字。

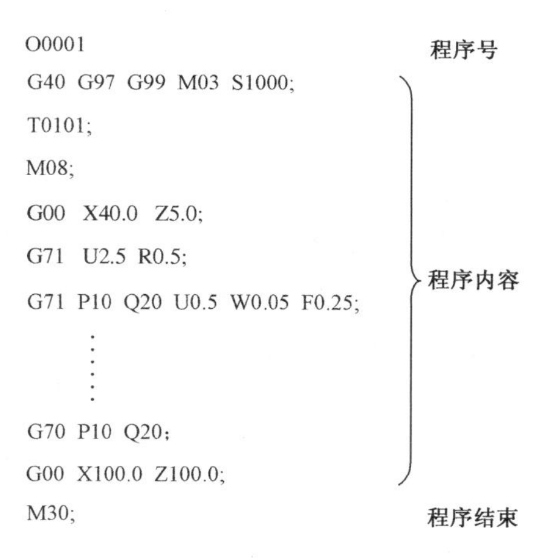

图 1-2-3　程序结构

(5)F、S、M 和 T 代码

F 代码控制进给速度(图 1-2-4),可以用每分钟进给量或每转进给量来表示。F 代码是模态代码,在随后的所有刀具运动程序中一直有效。在一个程序中可以根据需要改变进给速度。

S代码控制主轴转速(图1-2-5),S地址后面是所选择的转速值。S指令应和主轴旋转指令(M03或M04)一起使用。例如,“M03S1600;”表示设定数控机床主轴转速为 1 600 r/min。

M 代码叫作辅助功能字,用来控制机床的辅助功能,包括转动/停止主轴、启动/停止机床、开/关冷却液、换刀和返回程序头等功能。

T 代码用来指定刀具号(图 1-2-6)。

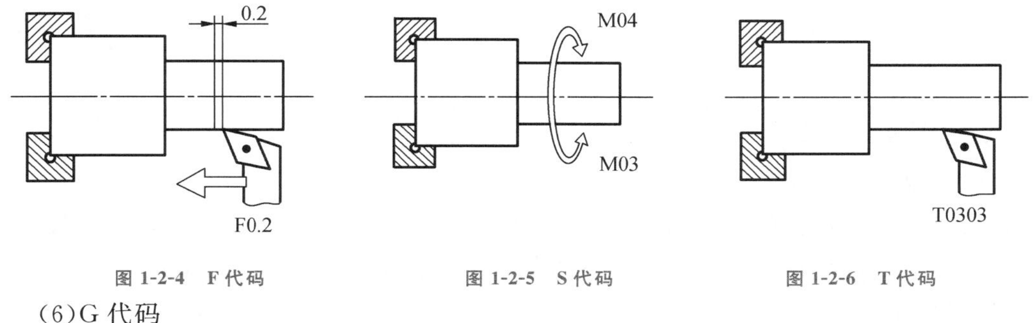

图 1-2-4　F 代码　　图 1-2-5　S 代码　　图 1-2-6　T 代码

(6)G 代码

G 代码也称为准备代码或字。准备代码由地址 G 及后面的数字组成,它规定了数控机床沿其编程轴运动的模式。G 代码通常放在程序段的开头。G 代码通常分为两类:模态 G 代码在随后的所有程序段一直有效,直至被同组的另一个 G 代码代替;非模态 G 代码仅在所出现的程序段内起作用。

2. 编程代码

(1)基本编程代码

◎ G00(快速定位)

G00 指令在工件坐标系中以快速移动速度移动刀具到达由绝对或增量指令指定的位置。在绝对指令中用终点坐标值编程,在增量指令中用刀具移动的距离编程。快速移动速度不能在地址 F 中被指定。指令格式为"G00 X(U)__ Z(W)__;"。终点坐标用 X(U)和 Z(W)指定,X、Z 表示绝对坐标值,U、W 表示增量坐标值。

◎ G01(直线插补)

G01 指令将刀具以 F 代码指定的进给速度沿直线移动到指定的位置。F 代码中指定的进给速度一直有效,直到指定新值。不必对每个程序段都指定 F 代码。G01 和 G00 指令都是模态 G 代码。指令格式为"G01 X(U)__ Z(W)__ F __;"。F 表示刀具进给速度。

◎ G02/G03(圆弧插补)

该指令使刀具沿圆弧运动。指令格式为"G02(G03) X(U)__ Z(W)__ R __(或 I __ K __) F __;"。*XZ* 平面上的顺时针(G02)和逆时针(G03)是在笛卡儿坐标系中从 *Y* 轴的正方向向负方向来观察 *XZ* 平面而定义的,如图 1-2-7 所示。

用 X(U)和 Z(W)指定圆弧终点。对于增量值,指定的是圆弧终点相对于圆弧起点的距离。

圆弧圆心由地址 I 和 K(分别对应于 *X* 和 *Z* 轴的坐标)指定。但是,I 和 K 后面跟的数值是从圆弧起点向圆心看的矢量分量,如图 1-2-8 所示。

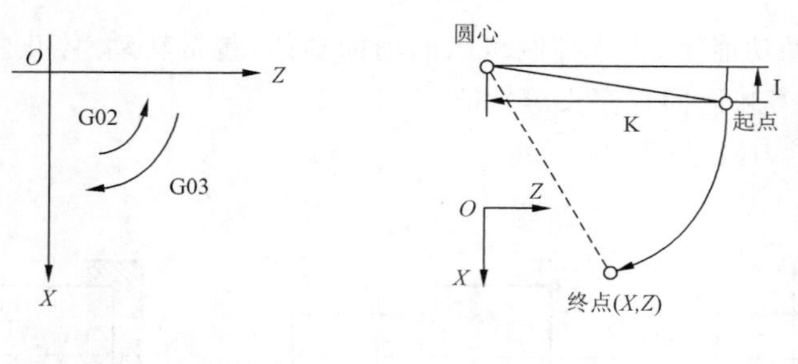

图 1-2-7 圆弧插补

图 1-2-8 I 和 K

◎ G04(暂停)

按指定的时间延迟执行下个程序段。指令格式为"G04 X __;"或"G04 U __;"或"G04 P __;"。X、U 和 P 表示指定时间(X、U 允许使用小数点,P 不允许使用小数点)。

(2)刀具补偿代码

大多数的数控系统使用三个 G 代码表示刀尖半径补偿。G41 用来设置刀尖半径左补偿,G42 用来设置刀尖半径右补偿,G40 用来取消刀尖半径补偿。

车削操作中如何使用 G41 和 G42,如图 1-2-9 所示。一旦刀尖半径补偿被正确设定,刀具将会一直保持在所有加工表面的左侧或右侧,直到 G40 代码取消刀尖半径补偿。

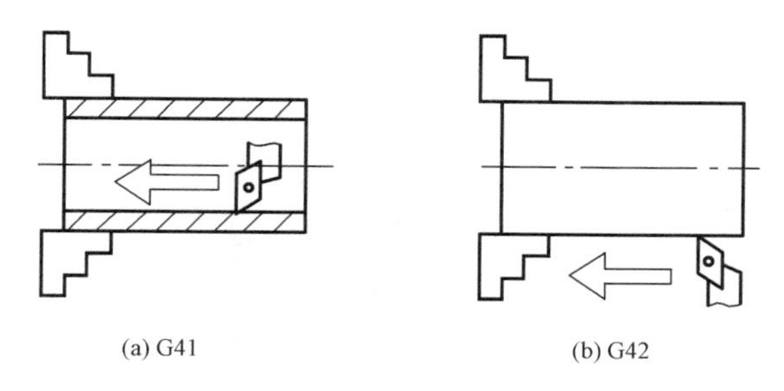

(a) G41　　(b) G42

图 1-2-9 G41 和 G42

(3)循环代码

◎ G71(粗车循环)

此循环指令能按照给定的精加工后的零件形状粗车去除多余材料。编程者需要做的是通过编制精加工刀具轨迹来定义零件形状,并且通过 PQ 程序段把指定刀具轨迹赋予 G71 指令。指令格式为:

G71 U(Δd) R(e);

G71 P(ns) Q(nf) U(Δu) W(Δw) F(f) S(s) T(t);

其中 Δd——切削深度(半径指定);

e——退刀量;

ns——精加工程序第一个程序段的顺序号;

nf——精加工程序最后一个程序段的顺序号;

Δu——X 方向精加工余量的距离和方向(直径值);

Δw——Z 方向精加工余量的距离和方向;

f、s、t——包含在 ns 到 nf 程序段中的任何 F、S 或 T 功能在循环中被忽略,而在 G71 程序段中的 F、S 或 T 功能有效。

◎ G73(仿形粗车循环)

通过这种切削循环,可以高效地切削已粗车成形、铸造成型或锻造成型等的工件。指令格式为:

G73 U(Δi) W(Δk) R(d);

G73 P(ns) Q(nf) U(Δu) W(Δw) F(f) S(s) T(t);

其中 Δi——X 方向退刀量的距离和方向(半径指定);

Δk——Z 方向退刀量的距离和方向;

d——均分次数。

其余参数的含义与 G71 相同。

◎ G70(精车循环)

用 G71 或 G73 粗切后,用 G70 指令完成精加工。指令格式为:

G70 P(ns) Q(nf);

其中,ns 和 nf 的含义与 G71 和 G73 指令中的相同。

3. 数控编程应用

零件图如图 1-2-10 所示,零件程序及其说明见表 1-2-1。

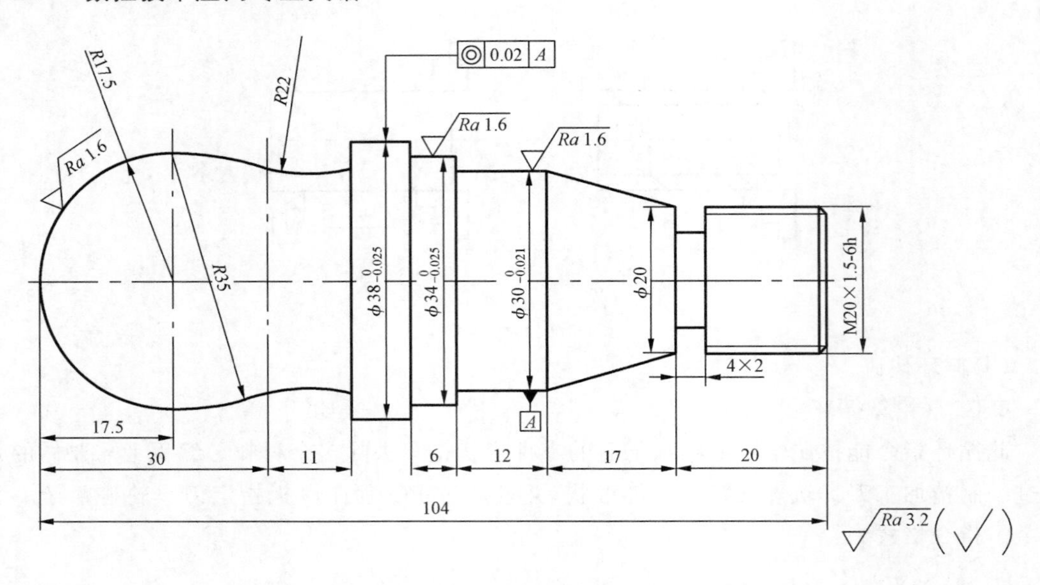

图 1-2-10 零件图

表 1-2-1 零件程序及其说明

零件程序	说明
O0001	0001 号程序
G40 G97 G99 M03 S600 F0.2;	取消刀尖半径补偿,设定主轴恒转速控制,主轴正转,转速为 600 r/min,进给量为 0.2 mm/r
T0101;	调用 1 号刀具及其刀具补偿
G00 G42 X40.0 Z5.0;	快速定位至点(*X*40.0,*Z*5.0),调用刀尖半径补偿
G71 U1.5 R0.5;	调用粗车循环
G71 P10 Q20 U0.5 W0.05;	
N10 G00 X0.0;	精加工轮廓
G01 Z0.0;	
X19.85 C1.5;	
Z−20.0;	
X20.0;	
X30.0 W−17.0;	
W−12.0;	
X34.0;	
W−6.0;	
X38.0;	
W−10.0;	
N20 X40.0;	
G00 G40 X100.0 Z100.0;	快速定位至点(*X*100.0,*Z*100.0),取消刀尖半径补偿
M03 S1000 T0101 F0.1;	主轴正转,转速为 1 000 r/min,调用 1 号刀具及其刀具补偿,进给量为 0.1 mm/r

续表

零件程序	说明
G00 G42 X40.0 Z5.0；	快速定位至点(X40.0,Z5.0)，调用刀尖半径补偿
G70 P10 Q20；	调用精车循环
G00 G40 X100.0 Z100.0；	快速定位至点(X100.0,Z100.0)，取消刀尖半径补偿
M03 S400 T0202 F0.08；	主轴正转，转速为 400 r/min，调用 2 号刀具(切槽刀)及其刀具补偿，进给量为0.08 mm/r
G00 Z−20.0；	快速定位至点(X100.0,Z−20.0)
X22.0；	快速定位至点(X22.0,Z−20.0)
G01 X16.0；	直线插补至点(X16.0,Z−20.0)
G04 X2.0；	调用暂停指令，暂停时间为 2 s
X22.0；	直线插补至点(X22.0,Z−20.0)
G00 X100.0 Z100.0；	快速定位至点(X100.0,Z100.0)
M03 S400 T0303；	主轴正转，转速为 400 r/min，调用 3 号刀具(螺纹刀)及其刀具补偿
G00 X21.0 Z5.0；	快速定位至点(X21.0,Z5.0)
G76 P020060 Q100 R0.1；	调用螺纹切削多重循环
G76 X18.05 Z−18.0 P975 Q400 F1.5；	
G00 X100.0 Z100.0；	快速定位至点(X100.0,Z100.0)
M30；	程序结束并返回程序头
O0002	0002 号程序
G40 G97 G99 M03 S600 F0.2；	取消刀尖半径补偿，设定主轴恒转速控制，主轴正转，转速为 600 r/min，进给量为 0.2 mm/r
T0101；	调用 1 号刀具及其刀具补偿
G00 G42 X40.0 Z5.0；	快速定位至点(X40.0,Z5.0)，调用刀尖半径补偿
G73 U20.0 W0 R10.0；	调用仿形粗车循环
G73 P10 Q20 U0.5 W0.05；	
N10 G00 X0.0；	精加工轮廓
G01 Z0.0；	
G03 X35.0 Z−17.5 R17.5；	
G03 X30.567 Z−30.0 R17.5；	
G02 X30.567 W−11.0 R22.0；	
N20 X40.0；	
G00 G40 X100.0 Z100.0；	快速定位至点(X100.0,Z100.0)，取消刀尖半径补偿
M03 S1000 T0101 F0.1；	主轴正转，转速为 1 000 r/min，调用 1 号刀具及其刀具补偿，进给量为 0.1 mm/r
G00 G42 X40.0 Z5.0；	快速定位至点(X40.0,Z5.0)，调用刀尖半径补偿
G70 P10 Q20；	调用精车循环
G00 G40 X100.0 Z100.0；	快速定位至点(X100.0,Z100.0)，取消刀尖半径补偿
M30；	程序结束并返回程序头

Task 3 CNC Lathe Operation

Speaking Activity

对话音频

A: Good afternoon, Mr. Li.

B: Good afternoon, Miss Wang. Could you help me?

A: What can I do for you?

B: Can you tell me how to input a new program?

A: First, switch the knob to "EDIT" mode on the panel and press the program key. Then input the program number with the address/number key. When the insert key is pressed, the program number appears on the screen. Finally, you can input the program content.

B: Which key can be used to change the feed rate?

A: You can rotate the feed rate knob to adjust the feed rate. You can select the feed rate you need. The feed rate increases if you rotate the feed rate knob clockwise. The feed rate decreases if you rotate the knob counterclockwise.

B: How to check the program?

A: Press the single block mode key during automatic operation. You can check the program in the single block mode by executing the program block by block.

B: Thank you very much. Bye-bye!

A: Bye!

Part A Technical Reading

1. CNC Lathe's Panel

(1) Control Panel (FANUC 0i System)

课文音频

The FANUC 0i system is one of the most common types of CNC system in the factory today. Usually, the control panel is located at the front of the machine and is equipped with a CRT and various keys, as shown in Fig. 1-3-1.

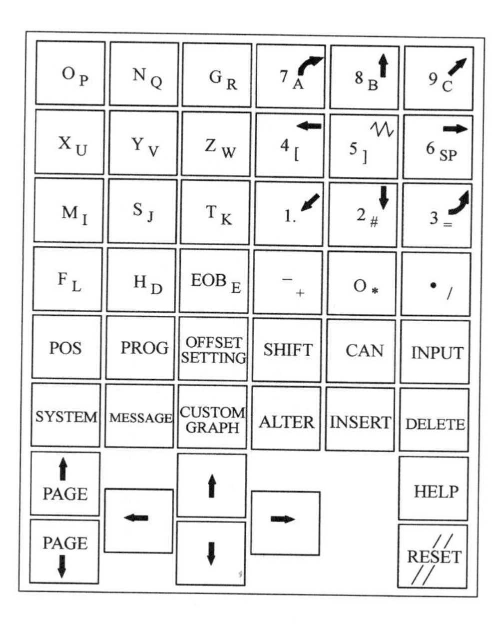

Fig. 1-3-1　Control Panel(FANUC 0i System)

The keys and their functions are shown in Chart 1-3-1.

Chart 1-3-1　　Keys and Their Functions of the Control Panel(FANUC 0i System)

Key	Name and Function
PROG	program key: display the program blocks of the selected part program on the screen
POS	position key: display different pages that list machine axis positions on the screen
OFFSET SETTING	menu/offset key: show offsets, you can input workpiece offset values and cutter compensation values
SYSTEM	system key: list machine parameters
MESSAGE	message key: show active alarm messages
CUSTOM GRAPH	graphics key: pressing the key once shows settings, pressing the key twice allows you to run a part program on screen
O P	address/number key: input the letters, numbers and so on
ALTER	alter key: the word will be changed by the new word
DELETE	delete key: the word will be deleted
INSERT	insert key: insert the characters, letters and numbers

continued

Key	Name and Function
CAN	cancel key: cancel the last character or symbol input to the buffer
EOB E	end of block key: at the end of each block, press the key ";" will appear
SHIFT	shift key: switch the characters
↑ PAGE, PAGE ↓	page change keys: paging forward or backward
←, ↑, ↓, →	cursor(arrow) keys: moving the cursor left, right, forward, backward
INPUT	input key: input workpiece offset values, cutter compensation values and so on
HELP	help key: indicate information for help
RESET	reset key: reset the CNC system or cancel an alarm
◄□□□□□►	soft keys: press the function keys first, then press the soft keys corresponding to the guide on the screen, you can enter the menu

(2) Machine Operation Panel

This task will introduce the machine operation panel(Fig. 1-3-2).

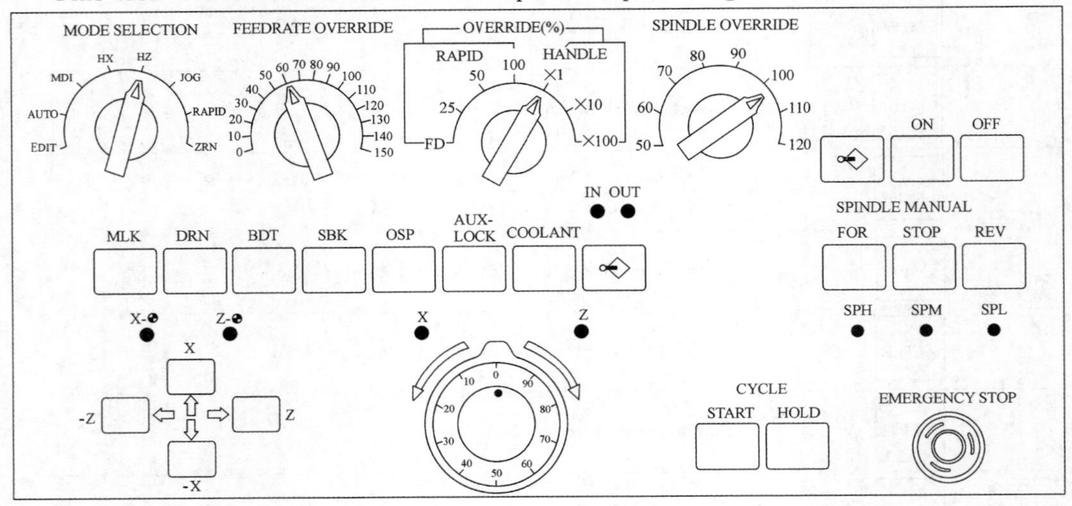

Fig. 1-3-2 Machine Operation Panel

The functions of the keys in the operation panel are shown in Chart 1-3-2.

Chart 1-3-2 Functions of the Keys in the Operation Panel

Key	Name and Function
MODE SELECTION (HX, HZ, MDI, JOG, AUTO, RAPID, EDIT, ZRN)	EDIT: edit mode enter a new program or modify the program
	AUTO: automatic mode make the machine run automatically
	MDI: manual data input mode enter and execute CNC command by manual
	HX: hand-wheel mode(select X axis) make the table move along X axis with the use of the hand-wheel
	HZ: hand-wheel mode(select Z axis) make the table move along Z axis with the use of the hand-wheel
	JOG: jog mode the selection of manual monotonous feed along the X or Z axis
	RAPID: rapid mode make the table move rapidly
	ZRN: reference point(zero return) mode key return to the zero position
0, 10, 20, 30, 40, 50, 60, 70, 80, 90, 100, 110, 120, 130, 140, 150	feed rate knob: adjust the axis feed speed
OVERRIDE(%) RAPID (FD, 25, 50, 100) HANDLE (×1, ×10, ×100)	RAPID: rapid feed rate override(F0, F25, F50, F100)
	HANDLE: hand-wheel feed rate override(×1, ×10, ×100)
SPINDLE OVERRIDE (50, 60, 70, 80, 90, 100, 110, 120)	spindle rate knob: adjust the spindle speed
ON OFF	ON: power on
	OFF: power off
SPINDLE MANUAL FOR STOP REV	FOR: spindle CW key spindle rotation clockwise
	STOP: spindle stop key spindle rotation stop
	REV: spindle CCW key spindle rotation counterclockwise

continued

Key	Name and Function
X Z	handle: the manual control of the movements of the table with the hand-wheel
X -Z Z -X	cursor keys: use these cursor keys to move the coordinate axis
CYCLE START	START: cycle start key make the program run automatically
CYCLE HOLD	HOLD: feed hold key all feed is interrupted, but the rotation is not affected
EMERGENCY STOP	emergency stop button: make the machine stop in the emergent state
MLK	machine lock key: lock the machine and dry run the program
DRN	dry run key: all the rapid work feed are changed to one chosen feed, used to check a new program on the machine without any work actually being performed by the tool
BDT	block skip key: the program block signed a tilted bar "/" in the beginning is neglected
SBK	single block mode key: only one block of information will be executed
OSP	option stop key: press the key M01 effective, press the key M01 again invalid
COOLANT	coolant key: turn on or cut off the cutting fluids

2. CNC Lathe Operation

(1) Manual Operation

◎ Manual Return of the Reference Point

- The mode selection switch is in the "ZRN" position.
- Use these cursor keys to move the coordinate axes.

◎ Manual Continuous Feed

- Switch to "JOG" mode.
- Rotate the feed rate knob to select the feed rate.
- According to the directions needed, press the corresponding cursor keys

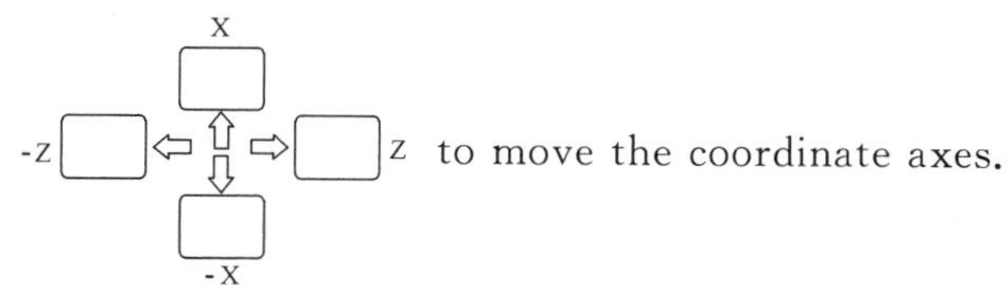

to move the coordinate axes.

Rotate the feed rate knob to adjust the feed rate. You can select the feed rate you need. The feed rate increases if you rotate the feed rate knob clockwise. The feed rate decreases if you rotate the knob counterclockwise.

◎ Handle Feed

- The mode selection switch is in the "HX" or "HZ" position.
- Select the feed rate override for the tool to be moved by rotating the switch

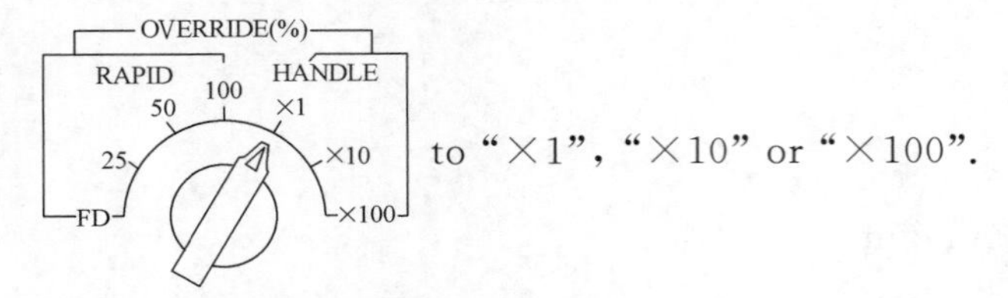

to "×1", "×10" or "×100".

(2) Inputting a Program

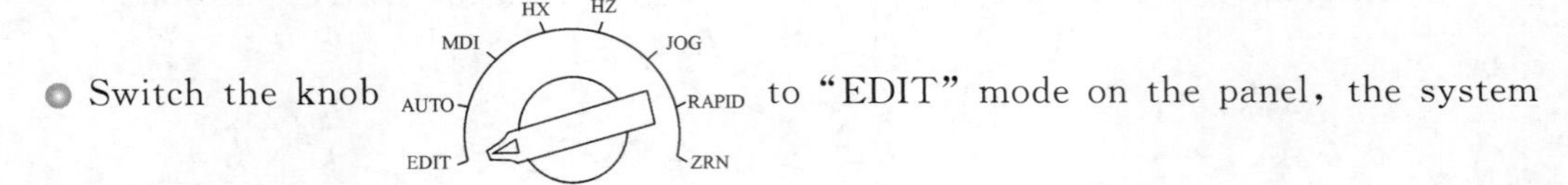

- Switch the knob to "EDIT" mode on the panel, the system will be in the edit mode.
- Press the program key [PROG] to display the program screen.
- Input the program number with the address/number key [O P].
- When the insert key [INSERT] is pressed, the program number appears on the screen.
- After that, you can input the program content. At the end of each line, press the key [EOB E]; ";" will appear on the screen. Then press the insert key [INSERT] again.

(3) Setting Offset Data

The procedure to set 1.0 as the *X* axis wear value for No. 3 tool is shown below:

- Press the offset key [OFFSET SETTING].
- Press the soft key [WEAR] and the tool wear interface is shown on the screen(Fig. 1-3-3).

OFFSET/WEAR　　O0001 N00000

NO.	X	Z.	R	T
W 001	0.000	1.000	0.000	0
W 002	1.486	-49.561	0.000	0
W 003	1.486	-49.561	0.000	0
W 004	1.486	0.000	0.000	0
W 005	1.486	-49.561	0.000	0
W 006	1.486	-49.561	0.000	0
W 007	1.486	-49.561	0.000	0
W 007	1.486	-49.561	0.000	0
W 008	1.486	-49.561	0.000	0

ACTUAL POSITION (RELATIVE)
0 101.000　　W 202.094

> _
MDI **** *** ***　　16:05:59
[WEAR][GEOM][WORK] [ ][(OPRT)]

Fig. 1-3-3　Tool Wear Interface

◎ Use the page change key [PAGE ↓] and the cursor keys [←] [↑] [↓] [→] to move the cursor to the position needed.

◎ Press the address/number key [O P] to enter "X1. 0", press the input key [INPUT], and thus the newly-input value is shown on the screen.

◎ If you want to revise the wear value, you just input the new value and press the input key [INPUT].

(4) MDI Operation

◎ The mode selection switch [MDI / HX / HZ / JOG / RAPID / ZRN / EDIT / AUTO] is in the "MDI" mode and the system will get into the MDI mode.

◎ Press the program key [PROG] and the MDI interface appears on the screen (Fig. 1-3-4).

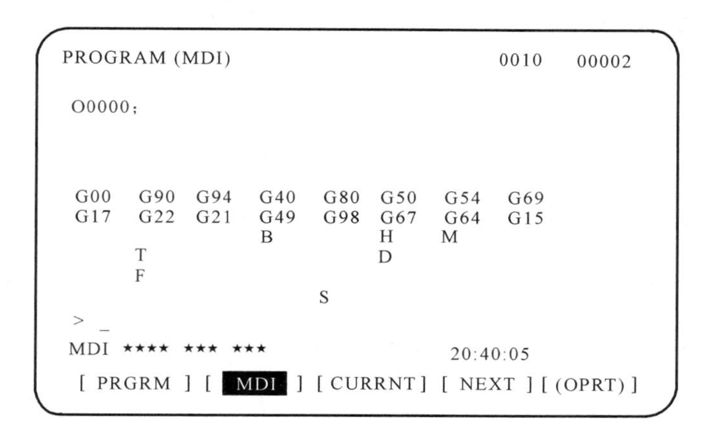

Fig. 1-3-4 MDI Interface

◎ Input one or more program blocks.

◎ Set the cursor at the beginning of the program's first block to execute the program.

◎ Press the cycle start key [START] and the prepared program will start.

(5) Automatic Operation

◎ Press the program key [PROG] and all the program files will appear on the screen. Select one of them and press the cursor key [↓] to open the program.

- Switch the knob to the "AUTO" mode and the CNC system will be in the automatic operation mode.
- Press the cycle start key START. When the indicator lights up, the program is to be executed.
- If you want to pause the program midway, you can press the cycle hold key HOLD and then the machine pauses. If you press the cycle start key START, the program will continue to be executed from the current block of the program.
- Press the reset key RESET and the program will stop. Press the cycle start key START and the program will be executed from the beginning.
- Press the dry run key DRN during automatic operation. The tool moves at the rapid feed rate specified in the parameter.
- Press the single block mode key SBK during automatic operation. The tool that stops after a single block in the program is executed. If you press the cycle start key START, the program will continue to be executed one block from the current block of the program. You can check the program in the single block mode by executing the program block by block.

New Words and Phrases

arrow /ˈærəʊ/	*n*. 箭，箭头记号
jog mode	点动模式
knob /nɒb/	*n*. 旋钮，按钮
increase /ɪnˈkriːs/	*n*. 增加，增大；*v*. 增加
decrease /dɪˈkriːs/	*n*. 减少，减少的量；*v*. 减少
retrieve /rɪˈtriːv/	*v*. 重新得到，调用

edit mode	编辑模式
wear /weə/	*n*. 磨耗
revise /rɪˈvaɪz/	*v*. 修改
interface /ˈɪntəfeɪs/	*n*. 界面
cursor key	光标键
manual data input	手动数据输入
feed hold	进给保持

Part B Practice Activity

Ⅰ. Match A with B.

A	B
手动操作	feed rate knob
连续进给	return to reference point
输入程序	manual operation
返回参考点	continuous feed
进给速度旋钮	input a program

Ⅱ. Answer the following questions briefly according to the text.

1. Which key can be used to move the axis?

____

2. Which key can be used to change the feed rate?

____

3. How do you increase the feed rate?

____

4. When inputting a program, which key do you press to enter ";" at the end of a line?

____

5. What is MDI mode?

____

6. Which key do you use to input a program number?

____

7. How do you revise the input value?

____

Ⅲ. Mark the following statements with T(true) or F(false).

(　　) 1. When you switch the knob to "JOG" mode, the system will be in the manual handle operation mode.

(　　) 2. Press the key REV, the spindle will rotate counterclockwise.

(　　) 3. Press the up X or the down -X arrow key, the $-Z$ axis indicator will light up.

(　　) 4. When the switch is in the "HZ" position and you rotate the hand wheel clockwise, the table will move in $+Z$ direction.

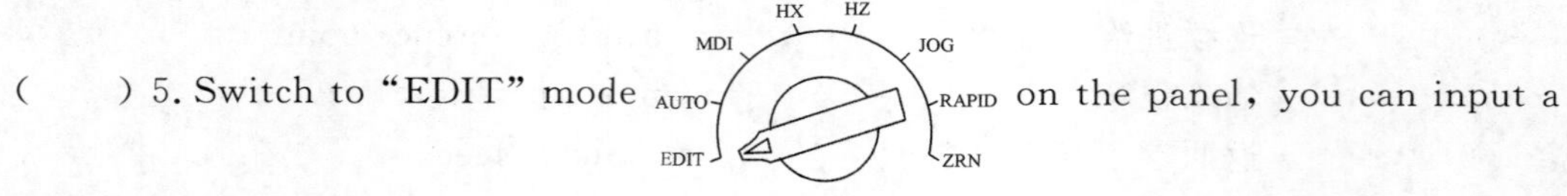

(　　) 5. Switch to "EDIT" mode on the panel, you can input a new program.

(　　) 6. The program number is not necessary when inputting a program.

(　　) 7. At the end of one line, press insert key INSERT, the cursor turns to next line.

(　　) 8. Tool wear value is the value to be used for the tool wear compensation.

Ⅳ. Fill in the blanks according to the text.

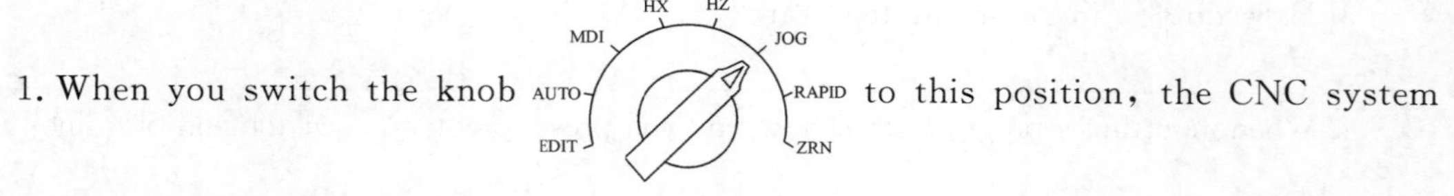

1. When you switch the knob to this position, the CNC system will switch into ________ mode.
2. If you rotate the feed rate knob clockwise, the feed rate ____________.
3. Press ________, the program screen appears.
4. We use the ________ to turn the pages.
5. If you want to modify the input value, you can input the new value and press the ____________.
6. The first step to set the tool wear value is to press the ________.

Professional Situation Simulation

Please write a composition in about 150 words according to the following key words.

Key Words: *manual operation*, *input a program*, *set offset data*, *MDI operation*, *automatic operation*

Work Sheet

....

....

....

....

....

Part C Broaden Your Horizon

Fig. 1-3-5 is the SIEMENS machine control panel.

拓展阅读音频

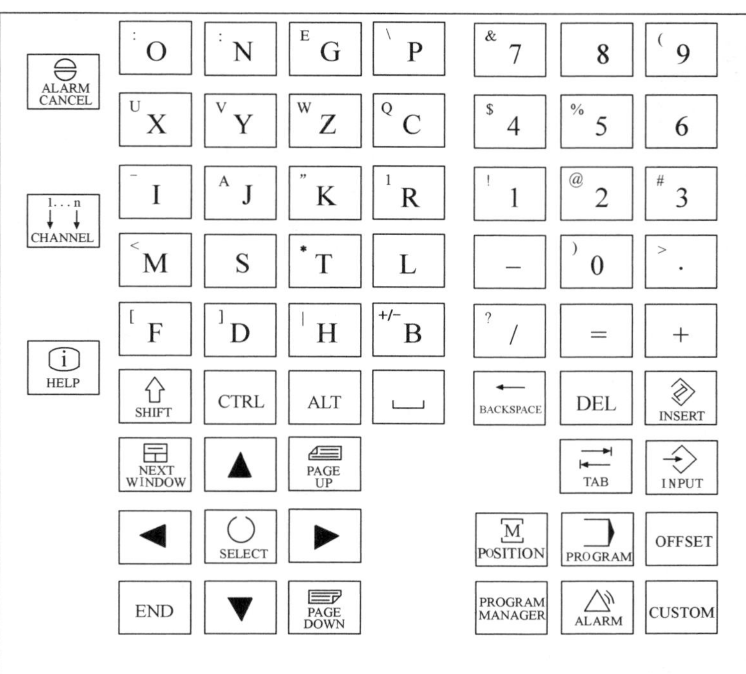

Fig. 1-3-5 SIEMENS Machine Control Panel

The keys and their names are shown in Chart 1-3-3.

Chart 1-3-3 Keys and Their Names

Key	Name	Key	Name
ALARM CANCEL	alarm acknowledgement key	(9	numeric keys double assignment on the shift level
W Z	alpha numeric keys double assignment on the shift level	1…n CHANNEL	channel switchover key
HELP	help key	SHIFT	shift key
CTRL	control key	ALT	alt key
␣	space key	BACKSPACE	deletion (backspace) key
DEL	deletion key	INSERT	insert key
NEXT WINDOW	not assigned key	▲	cursor key
PAGE UP	paging up key	TAB	tabulator key
INPUT	enter/input key	◀	cursor key
SELECT	selection key	▶	cursor key
M POSITION	position operating area key	PROGRAM	program operating area key
OFFSET	parameter operating area key	END	toggle key
▼	cursor key	PAGE DOWN	paging down key
PROGRAM MANAGER	program manager operating area key	ALARM	alarm operating area key
CUSTOM	system operating area key		

Translation

任务 3　数控车床操作

1. 数控车床面板

(1)控制面板(FANUC 0i 系统)

FANUC 0i 系统是当前工厂中最常用的数控系统。通常控制面板位于机床的前面,带有一个显示屏和各种按键,如图 1-3-1 所示。

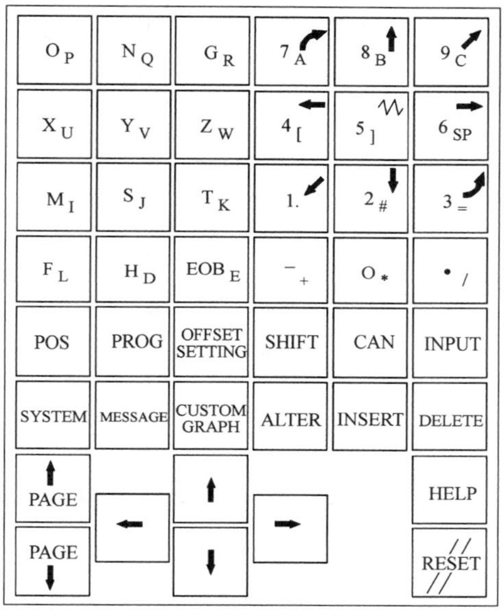

O P	N Q	G R	7 A	8 B	9 C
X U	Y V	Z W	4 [	5]	6 SP
M I	S J	T K	1 .	2 #	3 =
F L	H D	EOB E	− +	0 *	• /
POS	PROG	OFFSET SETTING	SHIFT	CAN	INPUT
SYSTEM	MESSAGE	CUSTOM GRAPH	ALTER	INSERT	DELETE
↑ PAGE	←	↑	→		HELP
PAGE ↓		↓			RESET

图 1-3-1　控制面板(FANUC 0i 系统)

按键及其功能见表 1-3-1。

表 1-3-1　控制面板上的按键及其功能(FANUC 0i 系统)

按键	名称与功能
PROG	程序键:在屏幕上显示所选择零件的程序
POS	位置键:在屏幕上显示机床坐标轴位置的不同界面
OFFSET SETTING	偏置键:显示偏移量,可以输入工件偏置值和刀具补偿值
SYSTEM	系统键:显示机床参数

续表

零件程序	注释
MESSAGE	信息键：显示激活的报警信息
CUSTOM GRAPH	图形显示键：按一次显示设置，再按此键将在屏幕上运行零件的程序
O_P	地址/数字键：输入地址、数字等
ALTER	替换键：旧字被新字代替
DELETE	删除键：字被删除
INSERT	插入键：插入字符、字母和数字
CAN	取消键：取消输入缓存器中的最后一个字符或符号
EOB_E	程序段结束键：在每一个程序段末尾按此键将出现“;”
SHIFT	切换键：切换字符
↑PAGE PAGE↓	翻页键：向前或向后翻页
↑ ← → ↓	光标（箭头）键：前、后、左、右移动光标
INPUT	输入键：输入工件偏置值和刀具补偿值等
HELP	帮助键：显示帮助信息

续表

零件程序	注释
RESET	复位键:数控系统复位或取消报警
◄□□□□□□►	软键:先按下功能键,然后按下与屏幕导向相应的软键,即可进入菜单

(2)机床操作面板

本任务介绍机床操作面板(图 1-3-2)。

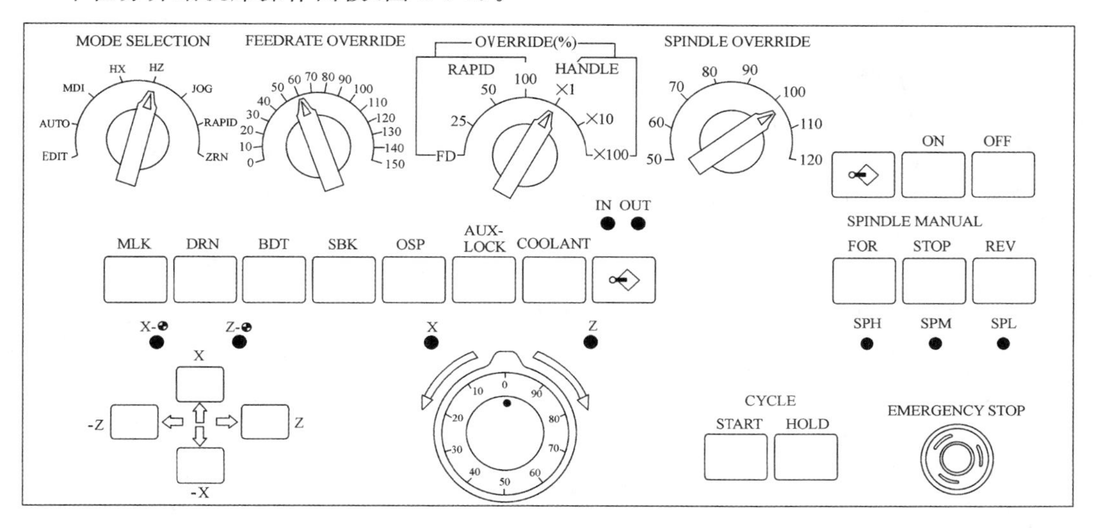

图 1-3-2　机床操作面板

操作面板上按键的功能见表 1-3-2。

表 1-3-2　　操作面板上按键的功能

按键	名称与功能
MODE SELECTION HX HZ MDI JOG AUTO RAPID EDIT ZRN	EDIT:编辑模式 输入或修改程序
	AUTO:自动模式 机床自动加工
	MDI:手动输入数据模式 手动输入和执行数控命令
	HX:手轮模式(选择 X 轴) 采用手轮方式使工作台沿 X 轴运动

续表

按键	名称与功能
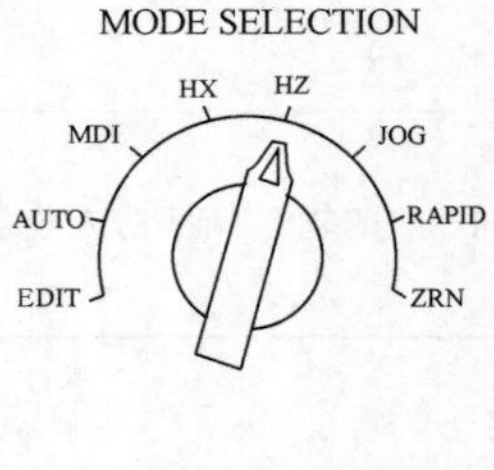 MODE SELECTION HX HZ MDI JOG AUTO RAPID EDIT ZRN	HZ:手轮模式(选择 *Z* 轴) 采用手轮方式使工作台沿 *Z* 轴运动
	JOG:点动方式 选择沿 *X* 或 *Z* 轴手动连续进给
	RAPID:快进模式 使工作台快速运动
	ZRN:参考点(返回零点)模式 返回零点位置
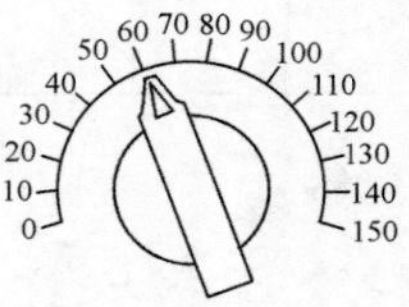 0 10 20 30 40 50 60 70 80 90 100 110 120 130 140 150	进给速度旋钮:调节进给轴的速度
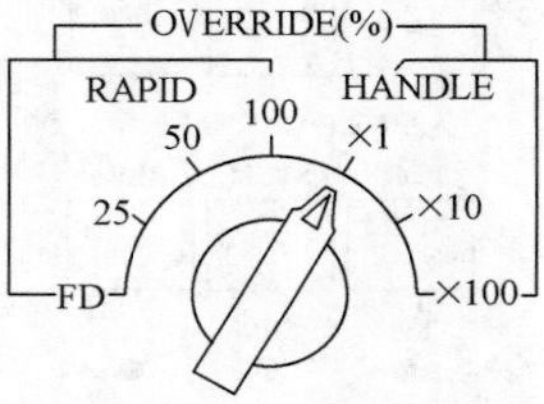 OVERRIDE(%) RAPID: FD 25 50 100 HANDLE: ×1 ×10 ×100	RAPID:进给倍率修调(F0、F25、F50、F100)
	HANDLE:手轮倍率修调(×1、×10、×100)
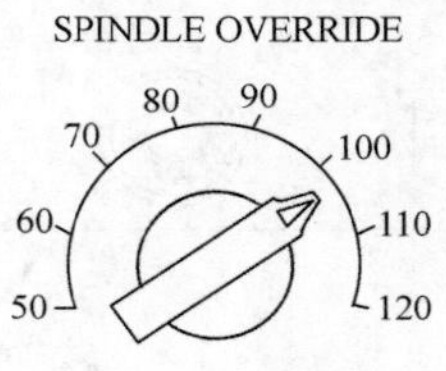 SPINDLE OVERRIDE 50 60 70 80 90 100 110 120	主轴速度旋钮:调节主轴转速
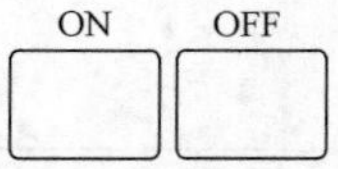 ON OFF	ON:接通电源
	OFF:关闭电源
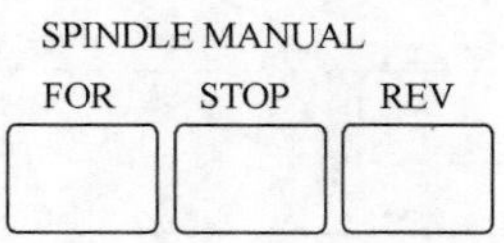 SPINDLE MANUAL FOR STOP REV	FOR:主轴正转键 主轴顺时针旋转
	STOP:主轴停止键 主轴停止旋转
	REV:主轴反转键 主轴逆时针旋转

续表

按键	名称与功能
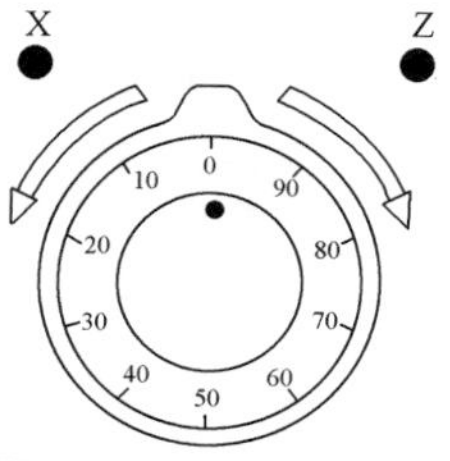 X Z 10 0 90 20 80 30 70 40 50 60	手轮：使用手轮手动控制工作台的移动
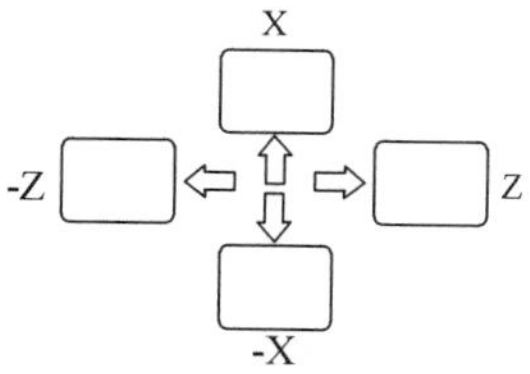 X -Z Z -X	光标键：用光标键移动坐标轴
CYCLE START	START：循环启动键 程序自动运行
CYCLE HOLD	HOLD：进给保持键 所有进给被中断，但不影响主轴旋转
EMERGENCY STOP 	紧急停止按钮：在紧急状态下使机床停止工作
MLK	机床锁住键：锁住机床，空运行程序
DRN	空运行键：所有快进和工作进给以一个选定的速度执行，用于在机床上刀具不切削工件的情况下检查新程序
BDT	程序段跳转键：开头有“/”的程序段被忽略
SBK	单段模式键：每次仅执行一个程序段内容
OSP	选择停止键：按下此键 M01 有效，再次按此键 M01 无效
COOLANT	冷却液键：打开或关闭冷却液

2. 数控车床操作

(1)手动操作

◎手动返回参考点

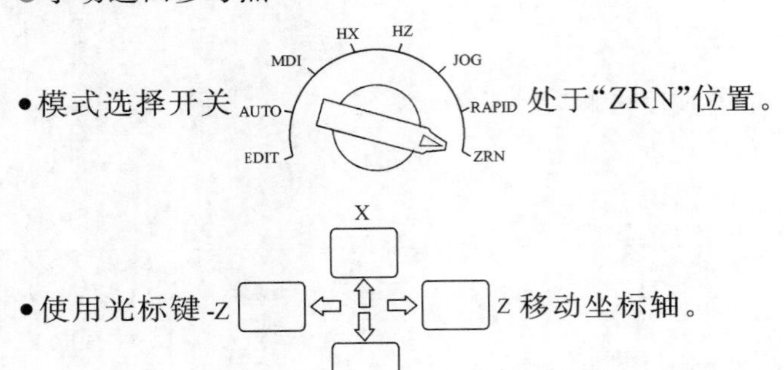

•模式选择开关处于“ZRN”位置。

•使用光标键-Z ⇦ ⇧ ⇨ Z 移动坐标轴。

◎手动连续进给

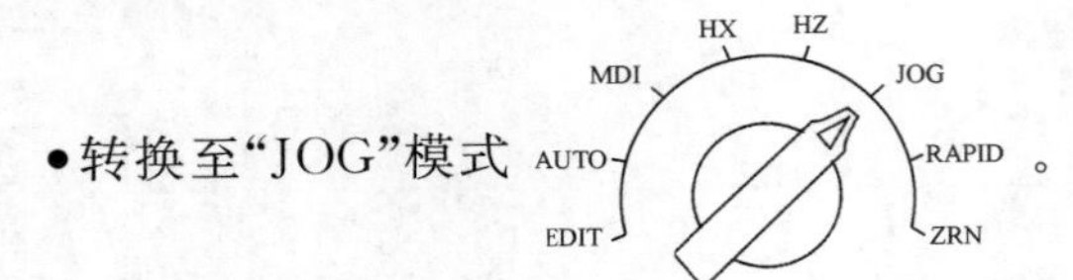

•转换至“JOG”模式。

•旋转进给速度旋钮 0 10 20 30 40 50 60 70 80 90 100 110 120 130 140 150 选择进给速度。

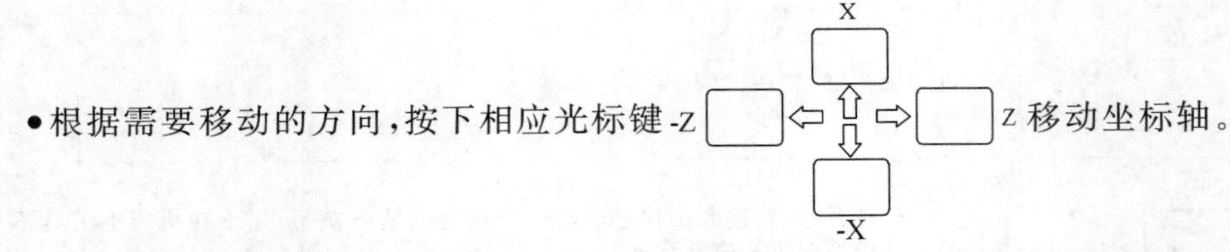

•根据需要移动的方向,按下相应光标键-Z ⇦ ⇧ ⇨ Z 移动坐标轴。

旋转进给速度旋钮调节进给速度,可以根据需要选择进给速度。顺时针旋转该旋钮,进给速度增大;逆时针旋转该旋钮,进给速度减小。

◎手轮进给

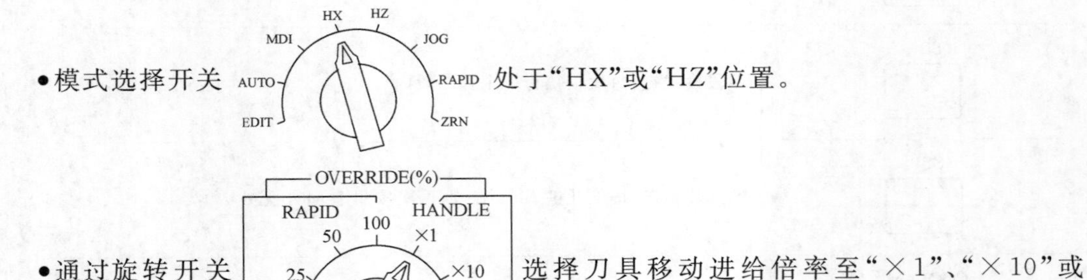

•模式选择开关处于“HX”或“HZ”位置。

•通过旋转开关选择刀具移动进给倍率至“×1”、“×10”或“×100”挡。

(2)输入程序

◎将面板上的旋钮转换至“EDIT”模式，系统则处于编辑模式。

◎按下程序键 PROG 显示程序屏幕。

◎使用地址/数字键 O P 输入程序号。

◎按下插入键 INSERT，程序号就会出现在屏幕上。

◎然后可以输入程序内容。在每行程序的结尾先按 EOB E 键，屏幕上将出现“;”，然后再按插入键 INSERT。

(3)设定偏置数据

设定 3 号刀的 *X* 轴磨耗值为 1.0 的步骤如下：

◎按下偏置键 OFFSET SETTING。

◎按下软键[WEAR]，刀具磨耗界面显示在屏幕上(图 1-3-3)。

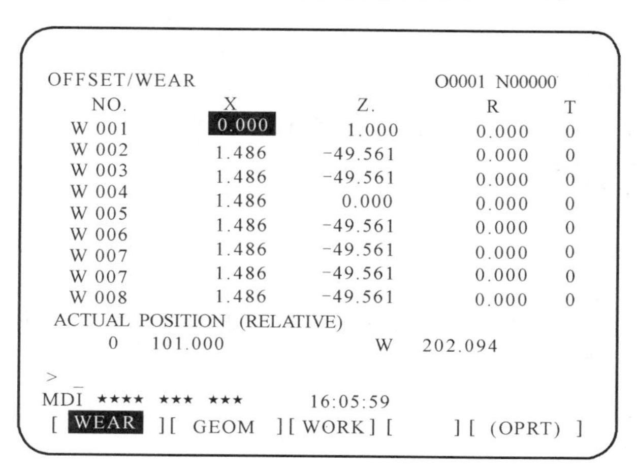

图 1-3-3 刀具磨耗界面

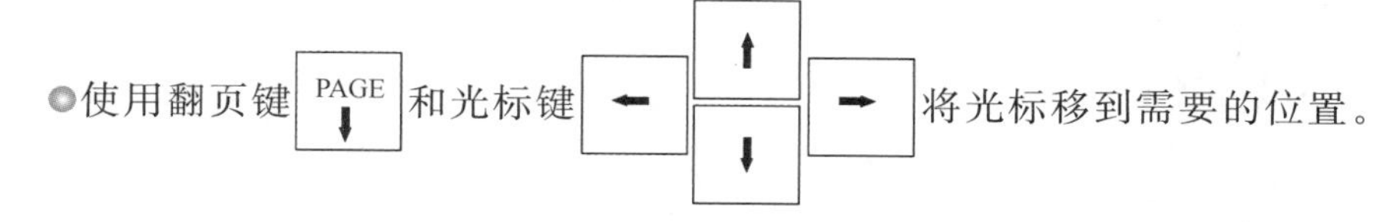

◎使用翻页键和光标键将光标移到需要的位置。

◎按地址/数字键 O P 输入“X1.0”，然后按输入键 INPUT ，这时新输入的数值就显示在屏幕上。

◎如果要修改输入的数值,可以直接输入新的数值,然后按输入键 INPUT 。

(4)MDI 操作

◎模式选择开关 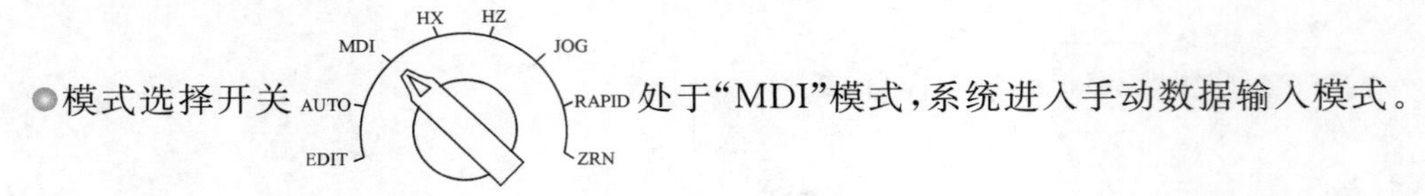

处于“MDI”模式,系统进入手动数据输入模式。

◎按下程序键 PROG ，MDI 界面出现在屏幕上(图 1-3-4)。

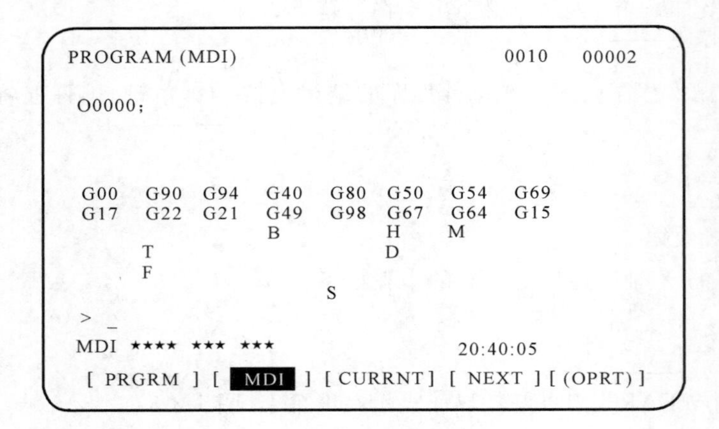

图 1-3-4 MDI 界面

◎输入一个或多个程序段。

◎将光标置于第一个程序段的开始,以执行程序。

◎按下循环启动键 START ,准备好的程序开始执行。

(5)自动操作

◎按下程序键 PROG ,所有的程序将出现在屏幕上。选定其中一个程序,然后按下光标键 ↓ 打开这个程序。

◎将旋钮 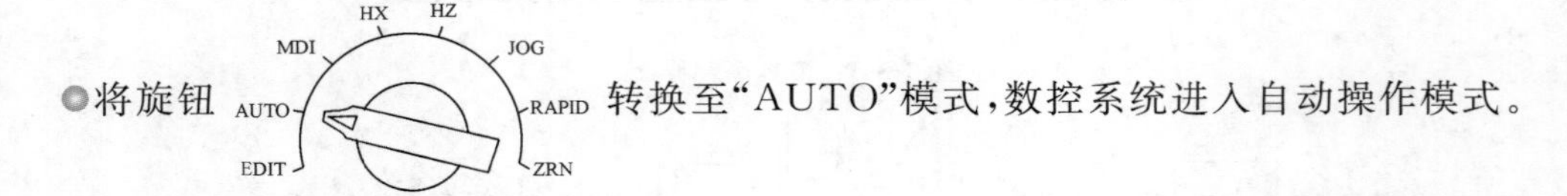

转换至“AUTO”模式,数控系统进入自动操作模式。

◎按循环启动键 START ,指示灯亮,该程序被执行。

◎如果要中途停止程序,可以按下进给保持键 HOLD ,则机床暂停运行。如果按下循环启动键 START ，则程序将从当前程序段被继续执行。

- 按下复位键 RESET 可以中断程序。按下循环启动键 START ，程序将从头开始被执行。
- 在自动操作过程中按下空运行键 DRN ，刀具以系统给定的参数快速移动。
- 在自动操作过程中按下单段模式键 SBK ，程序执行一段后刀具停止。如果按下循环启动键 START ，则将从当前程序段继续执行下一个程序段。通过一段一段地执行程序，可以在单段模式下检查程序。

学习启示

培养不断进取的工匠精神。“中国制造”在每一个工匠具体而微小的自我超越中走向更高层次的“中国创造”。

Module 2

CNC Machining Center Machining

Task 1 Basic Knowledge of CNC Machining Center

Speaking Activity

A: Good morning, Miss Wang.

B: Good morning, Mr. Li.

A: I haven't seen you for a long time. Where are you going?

B: I am going to the workshop.

A: Would you mind me asking you some questions?

B: No problem. What do you want to know?

A: What are the types of CNC machining center?

B: CNC machining centers are classified as vertical, horizontal and universal machining centers.

A: How can machining centers perform so many operations, such as milling, drilling or boring?

B: A machining center has many different cutters that are stored in the tool magazine and these cutters can be changed automatically.

A: Oh, I see. Then what are the components of a CNC machining center?

B: The main components of a CNC machining center are bed, saddle, column, table, servo system, spindle, tool changer and the MCU.

A: What is MCU?

B: MCU is a computer used to store and process the CNC programs entered.

A: I am very glad to know these. Thanks.

B: You are welcome.

Part A Technical Reading

1. The General Introduction About a CNC Machining Center

课文音频

CNC machining centers have been defined as multifunctional CNC machines with ATC (automatic tool changer) capabilities and tool magazine. Since their introduction in the late 1950s, they have become one of the most common of all cutting machines. Increased productivity and versatility are major advantages of CNC machining centers. We can perform drilling, reaming, boring, milling, contouring and threading operations on a CNC machining center. Most workpieces can be completed on a single CNC machining center, often with one setup.

CNC machining centers are classified as vertical, horizontal and universal machining centers.

For a vertical machining center (VMC), the best suitable types of workpiece are flat parts. The workpiece that requires machining two or more surfaces in a single setup is machined on a horizontal machining center (HMC), such as pump housing or boxy workpieces. Universal machining centers are equipped with both vertical and horizontal spindles. As they have multiple functions, they can machine all surfaces of a more complex workpiece.

CNC machining center innovations and developments have brought about the following improvements:

- Improved flexibility and reliability;
- Increased feed, spindle speed and overall machine's construction and rigidity;
- Reduced mounting, tool changing and other non-cutting time;
- Improved safety features.

2. Components of a CNC Machining Center

The main components of a CNC machining center are bed, saddle, column, table, servo system, spindle, tool changer and the machine control unit (Fig. 2-1-1).

The bed is usually made of high-quality cast iron. The bed supports all the components.

The table, which is mounted on the bed, provides the CNC machining center with the *X* axis linear movement.

The saddle, which is mounted on the hardened and grinding guide bed ways, provides the CNC machining center with the *Y* axis linear movement.

The column is mounted to the saddle. The column provides the CNC machining center with the *Z* axis linear movement.

The servo system, which consists of servo drive motor, ball screw and position feed-

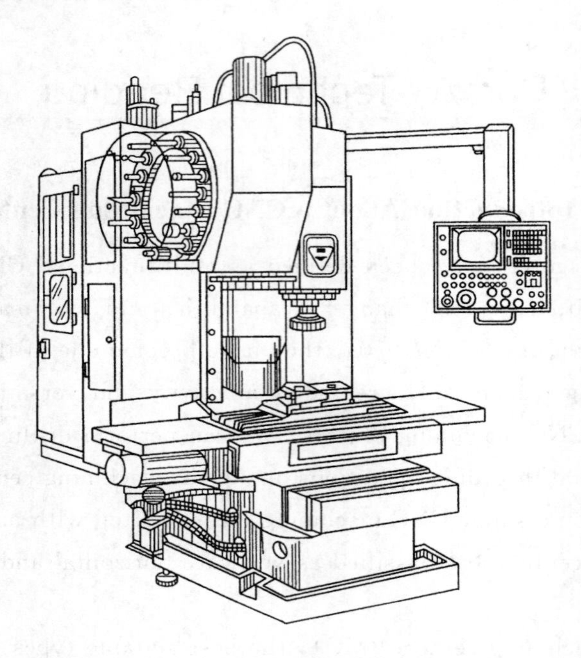

Fig. 2-1-1 Main Components of a CNC Machining Center

back encoder, provides fast, accurate movement and positioning of the *X*, *Y* or *Z* axis.

The tool changer is capable of storing a number of tools which can be automatically called by the program.

Machine control unit (MCU) is a computer used to store and process the CNC programs entered. New MCU is becoming more sophisticated, making machines more reliable and the entire machining operations less dependent on human skills.

New Words and Phrases

tool magazine	刀库
versatility /ˌvɜːsəˈtɪlətɪ/	*n.* 多用途，多功能
ream /riːm/	*vt.* 扩大，挖通，挤出
boring /ˈbɔːrɪŋ/	*adj.* 钻孔用的；*n.* 钻孔，钻屑
innovation /ˌɪnəˈveɪʃn/	*n.* 革新，改革，创新
flexibility /ˌfleksəˈbɪlətɪ/	*n.* 弹性，灵活性
saddle /ˈsædl/	*n.* 座板，滑动座架
feedback /ˈfiːdbæk/	*n.* 回应，反馈，回馈
sophisticated /səˈfɪstɪkeɪtɪd/	*adj.* 复杂的，精密的
pneumatic /njuːˈmætɪk/	*adj.* 充气的，气动的，装满空气的

Part B Practice Activity

Ⅰ. Match A with B.

A	B
horizontal CNC machining center	立式数控加工中心
tool magazine	线性运动
vertical CNC machining center	刀库
linear movement	机床控制单元
machine control unit	卧式数控加工中心

Ⅱ. Answer the following questions briefly according to the text.

1. What's the definition of a CNC machining center?

____

2. What are the main components of a CNC machining center?

____

3. What does the servo system consist of?

____

4. What's the meaning of MCU?

____

5. Which three kinds is a CNC machining center divided into?

____

6. What do you know about the vertical machining center?

____

Ⅲ. Mark the following statements with T(true) or F(false).

(　　) 1. Most workpieces can be completed with one setup on a single CNC machining center.

(　　) 2. CNC machining centers are classified as vertical and horizontal machining centers.

(　　) 3. The tool changer can store tools and change them automatically.

(　　) 4. CNC machining centers have automatic tool-changing capabilities and tool magazine.

(　　) 5. For a vertical machining center, the best types of workpiece are boxy workpiece.

Professional Situation Simulation

Please write a composition in about 150 words according to the following key words.

Key Words: *machining center*, *advantage*, *classification*, *component*

Work Sheet

....

....

....

....

....

....

....

....

....

....

....

Part C Broaden Your Horizon

A VMC has the spindle oriented in the vertical position. An automatic tool changer is mounted to the machine (usually on the left side) to allow tools to be loaded into the spindle without operator intervention. Basic VMC will allow three directions of motion. Some VMCs also have a rotary table or an indexer mounted on the table to allow a workpiece to be rotated during machining.

An HMC has the spindle oriented in the horizontal position. An automatic tool changer is equipped to allow tools to be automatically placed in the spindle. The table of most HMCs can rotate by either an indexer or a rotary axis. If the machine has a rotary axis, the rotation is called the *B* axis. An HMC is no more difficult to program than a VMC.

Translation

模块 2 数控加工中心加工

任务 1 认识数控加工中心

1. 数控加工中心概述

数控加工中心是具有自动换刀功能和刀库的多功能数控机床。数控加工中心自 20 世纪 50 年代后期出现以来,已经成为最常用的切削机床之一。数控加工中心的主要优点是提高了生产率并扩大了加工范围。在一台数控加工中心上可以实现钻孔、铰孔、镗孔、铣削、轮廓加工和攻螺纹等多种操作。大多数工件通常能在一台数控加工中心上用一次装夹完成全部加工。

数控加工中心可以分为立式、卧式和万能加工中心三种。

立式加工中心(VMC)最适合加工平板类零件。卧式加工中心(HMC)适合加工在一次装夹中有两个或更多表面被加工的零件,例如泵体或箱体类零件。万能加工中心同时具有垂直和水平主轴。由于它具有多种功能,故能加工更加复杂零件的所有表面。

数控加工中心的技术革新和发展已经带来了如下改进:

- 增强了柔性和可靠性;
- 提高了进给量、主轴转速和整个机床的结构与刚性;
- 减少了安装工件、换刀及其他非切削时间;
- 提高了安全性。

2. 数控加工中心的组成部件

数控加工中心的主要部件是床身、滑座、立柱、工作台、伺服系统、主轴、换刀装置和机床控制单元(图 2-1-1)。

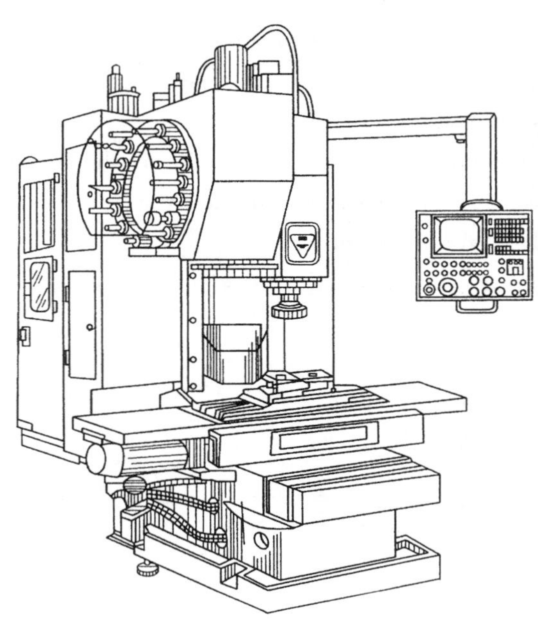

图 2-1-1 数控加工中心的主要部件

床身通常由优质铸铁制成，用来支撑所有部件。

工作台安装在床身上，为数控加工中心提供沿 X 轴的线性运动。

滑座安装在经过硬化和研磨处理的床身导轨上，为数控加工中心提供沿 Y 轴的线性运动。

立柱安装在滑座上，为数控加工中心提供沿 Z 轴的线性运动。

伺服系统包括伺服电动机、滚珠丝杠和位置反馈编码器，它可提供沿 X、Y 或 Z 轴的快速而准确的运动和定位。

换刀装置能够存储一定数量的刀具，零件加工程序可以自动调用这些刀具。

机床控制单元(MCU)是用于存储和处理所输入的数控程序的计算机。新型的 MCU 变得更加复杂，使机床可靠性更高，且整个加工操作更少地依赖于人的技巧。

Task 2 Manual Programming and Automatic Programming for CNC Machining Center

Speaking Activity

A: John, do you know something about the manual programming and the automatic programming?

B: Yes, a little.

A: What is the manual programming?

B: It means the programmer prepares the program.

A: How about the automatic programming?

B: The automatic programming is the use of computer special software to prepare CNC machining program. When the workpiece is complex in shape and tedious in calculation, it is necessary to use the automatic programming to improve work efficiency and ensure program quality.

A: In your opinion, it is not necessary for us to know the manual programming.

B: Not exactly. Successful use of the automatic programming requires understanding of the manual programming method.

A: That is to say, knowing the manual programming well will be the foundation to master the automatic programming.

B: Exactly.

A: Today I've got a lot. Thank you!

B: It is my pleasure to talk with you.

Part A Technical Reading

1. Manual Programming

(1) G94/G95 (Feed Rate)

Feed rate of linear interpolation, circular interpolation, etc. is commanded with numbers after the F code. G94 specifies the amount of feed of the tool per minute, and G95 specifies the amount of feed of the tool per spindle revolution, as shown in Fig. 2-2-1 and Fig. 2-2-2.

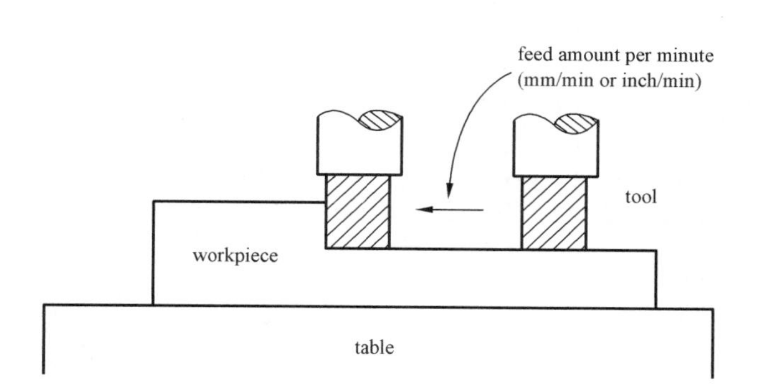

Fig. 2-2-1 Feed Amount per Minute

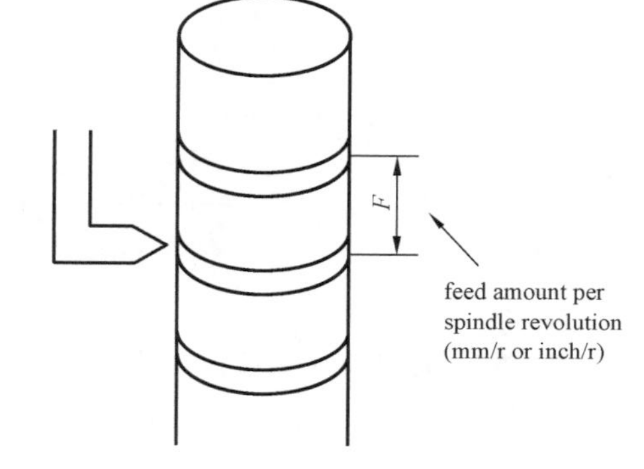

Fig. 2-2-2 Feed Amount per Spindle Revolution

(2) M98/M99 (Subprogram)

If a program contains a fixed sequence or frequently repeated pattern, this sequence or pattern can be stored as a subprogram in memory to simplify the program. The subprogram can be called from the main program, and the called subprogram can also call another subprogram.

◎ M98 (Subprogram Call)

The following is the format of program code:

M98 P□□□□○○○○;

In that, □□□□ refers to the times that the subprogram is called repeatedly. ○○○○ refers to subprogram number. When no repetition data is specified, the subprogram is called just once. A single subprogram call command can repeatedly call a subprogram up to 9 999 times.

◎ M99 (Subprogram End)

When M99 command is executed in a subprogram, the subprogram terminates and returns to the block after the calling block or the block with the sequence number specified by P. The following is the format of program code:

M99 (P○○○○);

(3) Tool Compensation

● G43/G44/G49(Tool Length Compensation)

Usually, several tools are used for machining one workpiece, especially in machining center. The tools have different tool lengths. It is very troublesome to change the program in accordance with the tools. Therefore, the length of each tool used should be measured in advance. By setting the difference between the length of the standard tool and the length of each tool in the CNC, machining can be performed without altering the program even when the tool is changed. This function is called tool length compensation, as shown in Fig. 2-2-3.

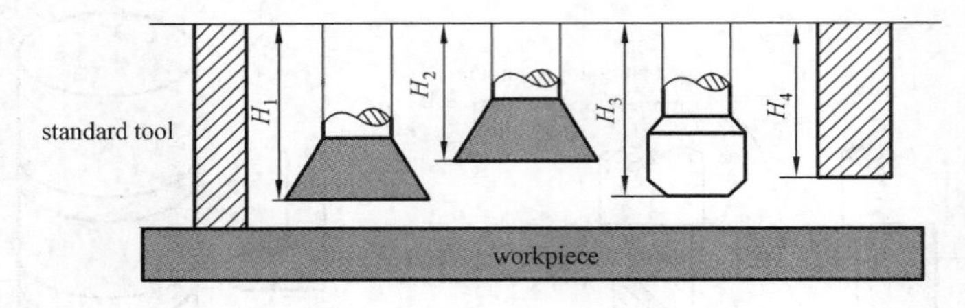

Fig. 2-2-3 Tool Length Compensation

● G41/G42/G40(Tool Radius Compensation)

When the tool is moving, the tool path can be shifted by the radius of the tool. This function is called tool radius compensation.

Most controls use three G codes for tool radius compensation: G41 is used to instate a tool radius left compensation (climb milling); G42 is used to instate a tool radius right compensation (conventional milling); G40 is used to cancel tool radius compensation. Additionally, many controls use a D word to specify the offset number.

There is a simplest way to determine whether to use G41 or G42 in the programming of machining center. If the tool is on the left side of the surface being machined, use G41; if on the right, use G42, as shown in Fig. 2-2-4.

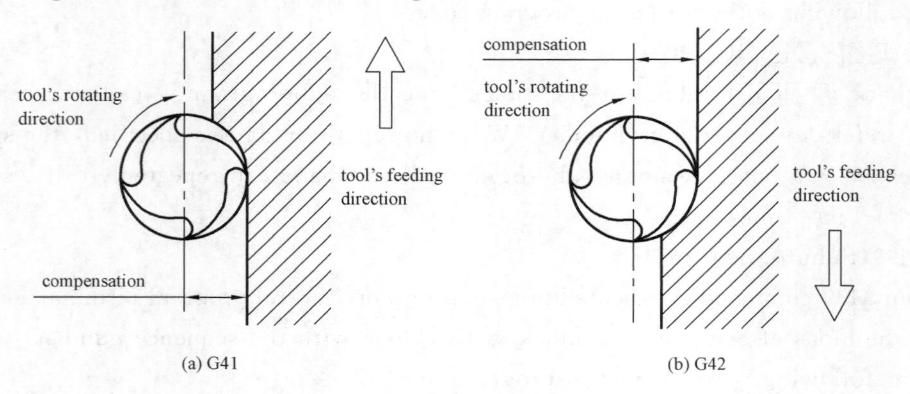

Fig. 2-2-4 G41 and G42

(4) Drilling Cycles

● G81(Drilling Cycle, Spot Drilling Cycle)

This cycle is used for normal drilling. Cutting feed is performed to the bottom of the hole. The tool is then retracted from the bottom of the hole in rapid traverse. When the tool reaches the bottom of the hole, the tool may return to point R or to the initial level. These operations are specified with G99 and G98. Generally, G99 is used for the first drilling operation and G98 is used for the last drilling operation, as shown in Fig. 2-2-5.

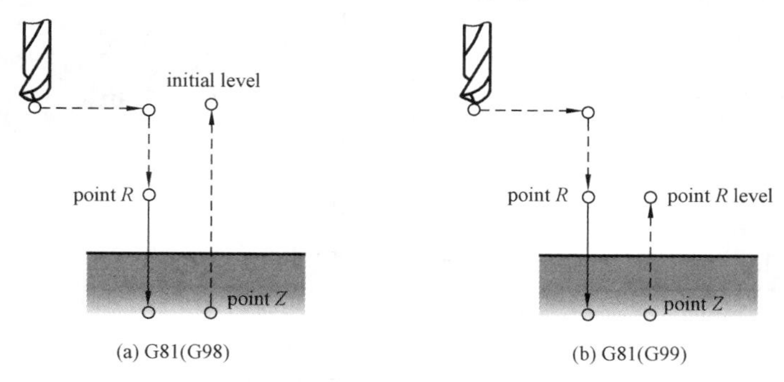

Fig. 2-2-5 G81

The following is the format of command:

G81 G98(G99) X_ Y_ Z_ R_ F_ K_;

In that X, Y—hole position data;

Z—the distance from point R to the bottom of the hole;

R—the distance from the initial level to point R level;

F—cutting feed rate;

K—numbers of repeat.

● G80(Canned Cycle Cancel)

All canned cycles are canceled by G80 command.

(5) G68/G69(Coordinate System Rotation)

A programmed shape can be rotated by using this function. Further, when there is a pattern comprising some identical shapes in the positions rotated from a shape, the time required for programming and the length of the program can be reduced by preparing a subprogram of the shape and calling it by coordinate rotation command(Fig. 2-2-6).

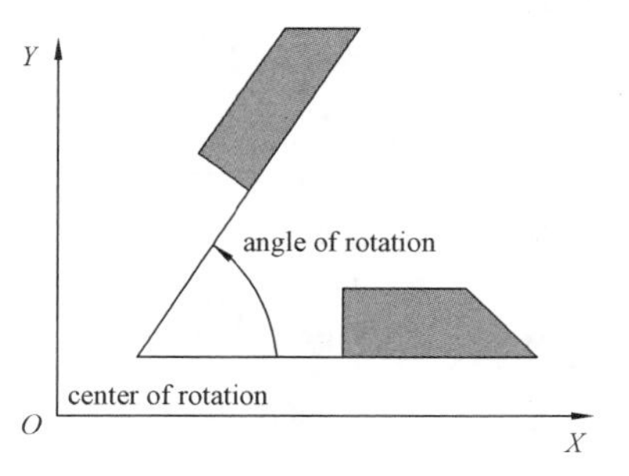

Fig. 2-2-6 Coordinate System Rotation

The format of command is:

G68—coordinate system rotation;

G69—coordinate system rotation cancel.

(6) Application of CNC Program

Part drawing is shown in Fig. 2-2-7. Part program and explanation are shown in Chart 2-2-1.

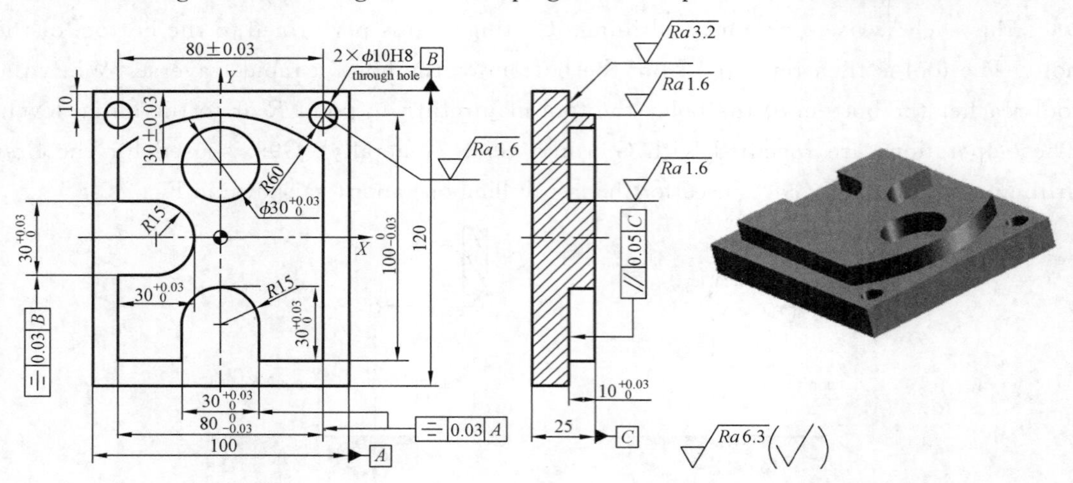

Fig. 2-2-7 Part Drawing

Chart 2-2-1 Part Program and Explanation

Program	Explanation
O0011	No. 0011 program (the outline machining)
N10 G90 G94 G21 G40 G54 F100;	setting: absolute coordinate program, feed per minute, selecting mm input, tool radius compensation cancelled, choosing No. 54 user's coordinate system, feed rate is 100 mm/min
N20 G91 G28 Z0;	return to reference point of *Z* axis
N30 M03 S600;	spindle CW at 600 r/min
N40 G90 G00 X−70.0 Y−80.0;	rapid positioning to point (*X*−70.0, *Y*−80.0)
N50 Z30.0 M08;	rapid positioning to *Z*30.0, cutting fluid on
N60 G01 Z−10.0;	linear interpolation to *Z*−10.0
N70 G41 G01 X−40.0 Y−80.0 D01;	tool radius compensation on, linear interpolation to point (*X*−40.0, *Y*−80.0)
N80 Y−15.0;	linear interpolation to point (*X*−40.0, *Y*−15.0)
N90 X−25.0;	linear interpolation to point (*X*−25.0, *Y*−15.0)
N100 G03 Y15.0 R15.0;	counterclockwise circular interpolation to point (*X*−25.0, *Y*15.0), the radius of circular arc is 15 mm
N110 G01 X−40.0;	linear interpolation to point (*X*−40.0, *Y*15.0)
N120 Y34.72;	linear interpolation to point (*X*−40.0, *Y*34.72)
N130 G02 X40.0 R60.0;	clockwise circular interpolation to point (*X*40.0, *Y*34.72), the radius of circular arc is 60 mm
N140 G01 Y−50.0;	linear interpolation to point (*X*40.0, *Y*−50.0)

continued

Program	Explanation
N150 X15.0;	linear interpolation to point(X15.0,Y−50.0)
N160 Y−35.0;	linear interpolation to point(X15.0,Y−35.0)
N170 G03 X−15.0 R15.0;	counterclockwise circular interpolation to point(X−15.0,Y−35.0), the radius of circular arc is 15 mm
N180 G01 Y−50.0;	linear interpolation to point(X−15.0,Y−50.0)
N190 X−60.0;	linear interpolation to point(X−60.0,Y−50.0)
N200 G40 G01 X−80.0 Y−80.0 M09;	tool radius compensation cancel, linear interpolation to point(X−80.0, Y−80.0), cutting fluid off
N210 G91 G28 Z0;	return to reference point of Z axis
N220 M30;	end program and reset
O0012	No. 0012 program(spot drilling)
N10 G90 G94 G21 G40 G54 F50;	setting: absolute coordinate program, feed per minute, selecting mm input, tool radius compensation cancel, choosing No. 54 user's coordinate system, feed rate is 50 mm/min
N20 G91 G28 Z0;	return to reference point of Z axis
N30 M03 S100;	spindle CW at 100 r/min
N40 G90 G00 Z50.0 M08;	rapid positioning to Z50.0, cutting fluid on
N50 G99 G81 X−40.0 Y50.0 Z−3.0 R−5.0 F50;	spot drilling cycle: tool rapidly moves to point(X−40.0, Y50.0), then rapidly moves to Z axis reference plane R−5.0, then feeds to the Z axis final depth Z−3.0 at speed 50 mm/min, at last rapidly back to the Z axis reference plane R−5.0
N60 G98 X40.0 Y50.0;	spot drilling the second hole, at last rapidly back to the initial level Z50.0
N70 X0 Y30.0;	spot drilling the third hole, at last rapidly back to the initial level Z50.0
N80 G80;	drilling cycle cancel
N90 G91 G28 Z0;	return to reference point of Z axis
N100 M30;	end program and reset
O0013	No. 0013 program(the inline machining)
N10 G90 G94 G21 G40 G54 F100;	setting: absolute coordinate program, feed per minute, selecting mm input, tool radius compensation cancel, choosing No. 54 user's coordinate system, feed rate is 100 mm/min
N20 G91 G28 Z0;	return to reference point of Z axis
N30 M03 S600;	spindle CW at 600 r/min
N40 G90 G00 X0 Y30.0;	rapid positioning to point(X0,Y30.0)
N50 Z20.0 M08;	rapid positioning to Z20.0, cutting fluid on
N60 G01 Z−10.0 F50;	linear interpolation to Z−10.0 at 50 mm/min
N70 G41 G01 X−5.0 Y30.0 D01 F100;	tool radius compensation on, linear interpolation to point(X−5.0,Y30.0) at 100 mm/min

continued

program	explanation
N80 G03 X15.0 R5.0;	counterclockwise circular interpolation to point (X15.0, Y30.0), the radius of circular arc is 5 mm
N90 G03 I15.0;	(machining the full circle) counterclockwise circular interpolation, X and Y vector components of arc center viewed from the start point are 15.0 and 0, respectively
N100 G40 G01 X0 Y30.0;	tool radius compensation cancel, linear interpolation to point(X0, Y30.0)
N110 G91 G28 Z0;	return to reference point of Z axis
N120 M30;	end program and reset
O0014	No. 0014 program(reaming)
N10 G90 G94 G21 G40 G54 F50;	setting: absolute coordinate program, feed per minute, selecting mm input, tool radius compensation cancel, choosing No. 54 user's coordinate system, feed rate is 50 mm/min
N20 G91 G28 Z0;	return to reference point of Z axis
N30 M03 S200;	spindle CW at 200 r/min
N40 G90 G00 Z50.0 M08;	rapid positioning to Z50.0, cutting fluid on
N50 G85 X−40.0 Y50.0 Z−30.0 R−5.0 F50;	boring cycle: tool rapidly moves to point (X−40.0, Y50.0), then rapidly moves to Z axis reference plane R−5.0, then feeds to the Z axis final depth Z−30.0 at speed 50 mm/min, at last rapidly back to the Z axis reference plane R−5.0
N60 X40.0 Y50.0;	reaming the second hole
N70 G80;	drilling cycle cancel
N80 G91 G28 Z0;	return to reference point of Z axis
N90 M30;	end program and reset

2. Automatic Programming

There are many types of automatic programming softwares such as Mastercam, UG, Pro/ENGINEER, etc.. Mastercam X is the widely used software for CNC automatic programming. Its main functions are as follows: create the geometry modeling; create the tool path; verify the cutting process; create the G codes.

(1) Create the Geometry Modeling

You can create the geometry modeling in one of two ways:

- By using the graphical interface design provided by Mastercam X;
- By making the design in CAD software, e. g. UG, Pro/ENGINEER, SolidWorks then saving it in a format that Mastercam X can import.

In this task we will use Mastercam X software to create the geometry modeling, as shown in Fig. 2-2-8.

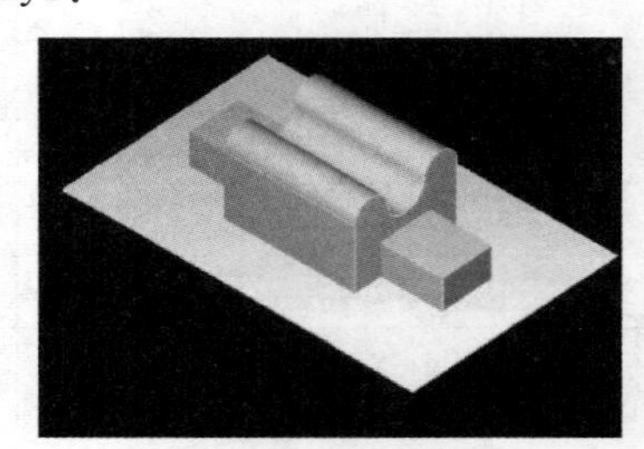

Fig. 2-2-8 Geometry Modeling

(2) Create the Tool Path

◎ Select the Machine Type

Click [Machine Type] → [Mill], select the machine type.

◎ Select an Existing Tool You May Want to Use

Click [Tool Paths] → [Tool Manager], in the "Tool Manager" dialog box, all available tools are displayed. You can search for a particular type by scrolling down the list or by using the "Filter" button and entering the search information.

◎ Create the Tool Path

- Click the main menu [Tool Paths] → [Surface Rough] → [Rough Parallel Tool Path] → dialog box will pop up → select "Boss" → click ✓.
- Select the part surfaces to be machined: click on the surfaces that you wish to machine on the part. Make sure that all the surfaces have been selected then click ✓.
- Set parameters of rough machining parallel milling.

(3) Verify the Cutting Process

Before a part is machined, the CAM model needs to be verified that the part program is correct. The purposes of verification are:

◎ To detect geometric errors of the tool path;

◎ To detect potential tool interferences;

◎ To detect erroneous cutting conditions.

To start verification, select one or more operations in the [Tool Path Manager List], then click the verify button, use the control buttons located at the top of the "Verify" dialog box to run simulation. The result of verification is shown in Fig. 2-2-9.

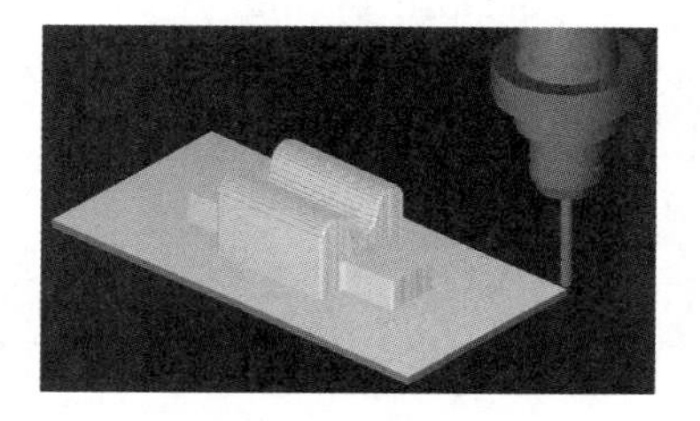

Fig. 2-2-9 The Result of Verification

(4) Create the G Codes

Different CNC machines use slightly different versions of G codes. The conversion of the machining data to the G codes for a particular CNC machine is called post-processing. The format of the G code is stored in different post-processing files and the system can use whichever post-processing format you select.

From the left panel in the [Tool Paths Tab] select the icon G1 → the "Post-processing" dialog box pops up → click ✓ → input the file name. Once the G codes are generated, we can see the G code files and modify them if required.

New Words and Phrases

tool length compensation	刀具长度补偿
tool radius compensation	刀具半径补偿
drilling cycle	钻孔循环
initial level	初始平面
the outline machining	外轮廓加工
automatic programming	自动编程
software /ˈsɔftwɛə/	*n.* 软件
tool path	刀具路径
scroll /skrəʊl/	*v.* 滚动
filter /ˈfɪltə/	*n.* 过滤器
erroneous /ɪˈrəʊniəs/	*adj.* 错误的,不正确的
verification /ˌverɪfɪˈkeɪʃn/	*n.* 验证,确认,证明,核实
slightly /ˈslaɪtli/	*adv.* 轻微地,稍微地,微小地
post-processing	后处理
tab /tæb/	*n.* 标签
icon /ˈaɪkɒn/	*n.* 图标

Part B Practice Activity

Ⅰ. Match A with B.

A	B
刀具长度补偿	surface rough
调用子程序	geometry modeling
刀具路径	post-processing
G 代码	subprogram call
自动编程	tool path
后处理	tool length compensation
表面粗加工	drilling cycle
钻孔循环	G code
几何模型	automatic programming

Ⅱ. Answer the following questions briefly according to the text.

1. What are the differences between program codes G94 and G95?

____

____

2. When had we better use subprogram?

____

____

3. What are the advantages of using tool length compensation?

____

____

4. What's the determinant factor to use G41 or G42?

____

5. Which ways can be used to create the geometry modeling?

____

____

6. How do we look for an existing tool that we want to use?

____

____

7. What is the purpose of verifying the cutting process?

____

____

____

Ⅲ. Mark the following statements with T(true) or F(false).

( ) 1. When we use program code M98, we must specify repetition data.

( ) 2. When M99 code is executed in a subprogram, the subprogram only returns to the block after the calling block.

( ) 3. Mastercam X is the only used software for CNC automatic programming.

( ) 4. Before a part is machined, the CAM model needs to be verified that the part program is correct.

( ) 5. Different CNC machines use the same versions of G codes.

Ⅳ. Fill in the blanks according to the text.

1. When a subprogram is used, there must be a ________ or frequently repeated pattern in a program.

2. When several tools are used for machining one workpiece in one setup, we usually use ________ to simplify the program.

3. The operation the tool returns to the ________ is specified with G98 in drilling cycles.

4. You can search for a particular type by scrolling down the list or by using the

______________ button and entering the search information.

5. The conversion of the machining data to the G codes specific for a particular CNC machine is called ______________.

Professional Situation Simulation

Please write a composition in about 150 words according to the following key words.

Key Words: *manual programming*, *subprogram*, *tool length compensation*, *tool radius compensation*, *automatic programming*, *create the geometry modeling*, *create tool path*, *verify the cutting process*, *create G codes*

Work Sheet

Part C Broaden Your Horizon

Direct editing of the program with the codes provided by the CNC machine is called manual programming. For the manual programming method, the efficiency and productivity of the workpiece program depend on the programmer's ability. Therefore, knowledge about process planning, machining theory, G codes and complex computations for tool path are necessary for a good programmer. In addition, it is almost impossible to create a workpiece program for machining 2.5D or 3D shape using the manual programming.

The automatic programming method, where a computer is used, was developed to overcome the above-mentioned problems with the manual programming method. The automatic programming method makes it easy to machine the workpiece with complex or 3D shapes. It also makes it possible to generate the large workpiece programs in a short time. In addition, with computer simulation, it makes it possible to detect and modify machining errors before actual machining begins.

Translation

任务 2 数控加工中心手工编程与自动编程

1. 手工编程

(1)G94/G95(进给速度)

直线插补、圆弧插补等的进给速度是由 F 代码后面的数值指定的。G94 指定每分钟的刀具进给量,G95 指定主轴每转的刀具进给量,如图 2-2-1 和图 2-2-2 所示。

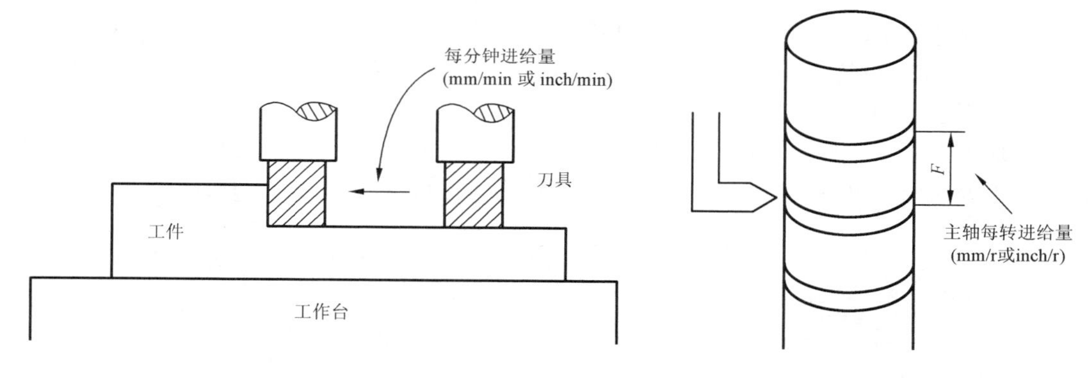

图 2-2-1 每分钟进给量　　图 2-2-2 主轴每转进给量

(2)M98/M99(子程序)

如果程序包含固定的顺序或多次重复的模式,则可将这种顺序或模式编成子程序储存在存储器中以简化编程。子程序可以由主程序调用,被调用的子程序也可以调用另一个子程序。

◎ M98(子程序调用)

程序代码的格式为:

M98 P□□□□○○○○;

其中,□□□□指子程序被重复调用的次数,○○○○指子程序号。当重复次数未被指定时,子程序仅被调用一次。单个子程序调用指令能重复调用一个子程序高达 9 999 次。

◎ M99(子程序结束)

在子程序中执行 M99 指令时,子程序就结束了,并且返回到调用程序段之后的程序段,或返回到由地址 P 指定顺序号的程序段。程序代码的格式为:

M99 (P○○○○);

(3)刀具补偿

◎ G43/G44/G49(刀具长度补偿)

通常加工一个零件要用多把刀具,尤其是使用加工中心加工零件时。刀具有不同的长度,根据使用的刀具去更换程序是相当麻烦的。因此,应事先测量所使用的每把刀具的长度。通过在数控机床中设定标准刀具的长度和每把刀具的长度之间的差,即使刀具改变了,也无须更换程序便可实现加工。该功能称为刀具长度补偿,如图 2-2-3 所示。

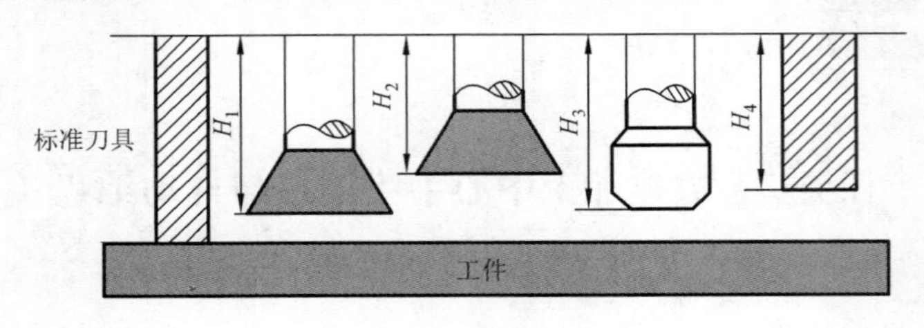

图 2-2-3 刀具长度补偿

◎ G41/G42/G40(刀具半径补偿)

当刀具移动时,刀具路径可以根据刀具半径进行偏移,该功能称为刀具半径补偿。

大多数控制系统使用三个 G 代码设置刀具半径补偿。G41 用来设置刀具半径左补偿(顺铣),G42 用来设置刀具半径右补偿(逆铣),G40 用来取消刀具半径补偿。此外,多种控制系统使用字符 D 来指定偏移量。

在加工中心上编程时,有一个最简单的方法用于判定是调用 G41 还是 G42。如果刀具在被加工表面的左侧,就调用 G41;如果在右侧,就调用 G42,如图 2-2-4 所示。

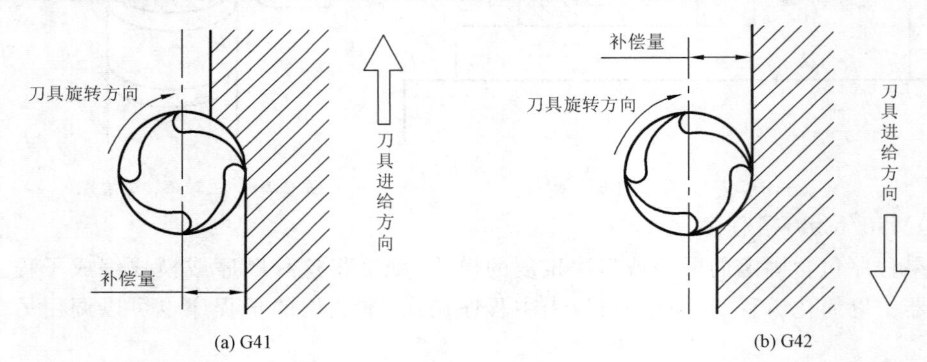

图 2-2-4 G41 与 G42

(4)钻孔循环

◎ G81(钻孔循环,钻中心孔循环)

该循环用于正常钻孔,切削进给执行到孔底,然后刀具从孔底快速移动退回。当刀具到达孔底后,刀具可以返回到 R 点平面或初始位置平面,由 G99 和 G98 指定。一般情况下,G99 用于第一次钻孔,G98 用于最后一次钻孔,如图 2-2-5 所示。

指令格式为:

G81 G98(G99) X_ Y_ Z_ R_ F_ K_;

其中 X、Y——孔位数据;

Z——从 R 点到孔底的距离;

R——初始平面与 R 点平面之间的距离;

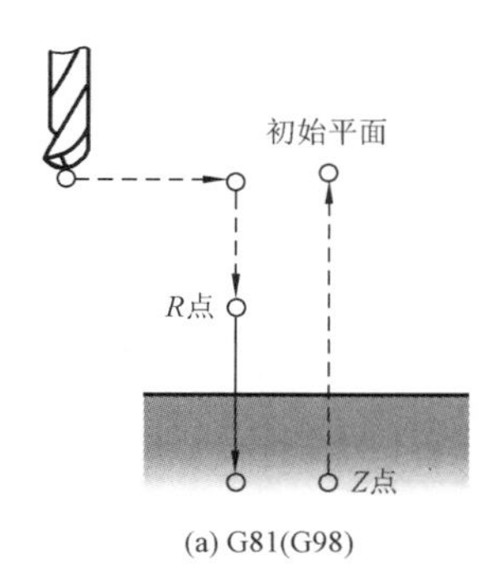

(a) G81(G98)

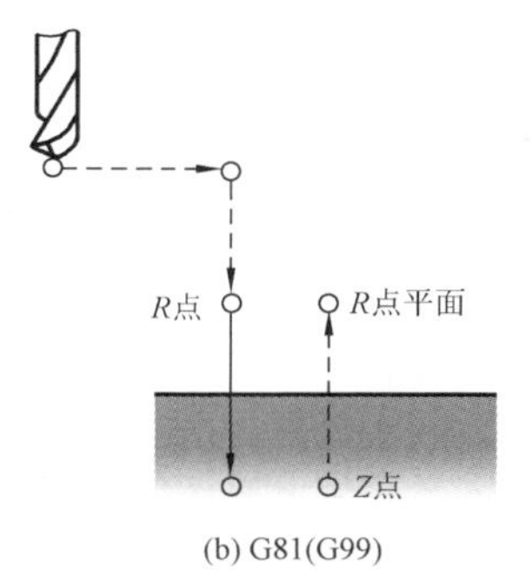

(b) G81(G99)

图 2-2-5　G81

F——切削进给速度；

K——重复次数。

◎ G80(取消固定循环)

G80 指令取消所有的固定循环。

(5)G68/G69(坐标系旋转)

使用该功能编程，形状能被旋转。此外，如果工件形状由许多相同的图形组成，则可将图形单元编成子程序，然后用坐标旋转指令调用，这样既可以省时，又可以使程序长度缩短(图 2-2-6)。

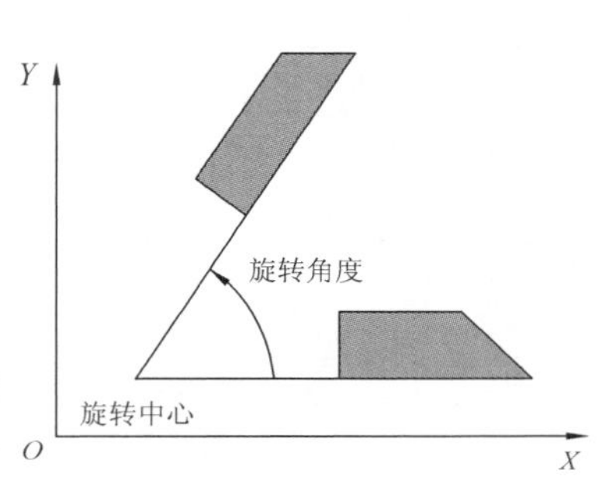

图 2-2-6　坐标系旋转

指令格式为：

G68——坐标系旋转；

G69——坐标系旋转取消。

(6)数控编程应用

零件图如图 2-2-7 所示，零件程序及其说明见表 2-2-1。

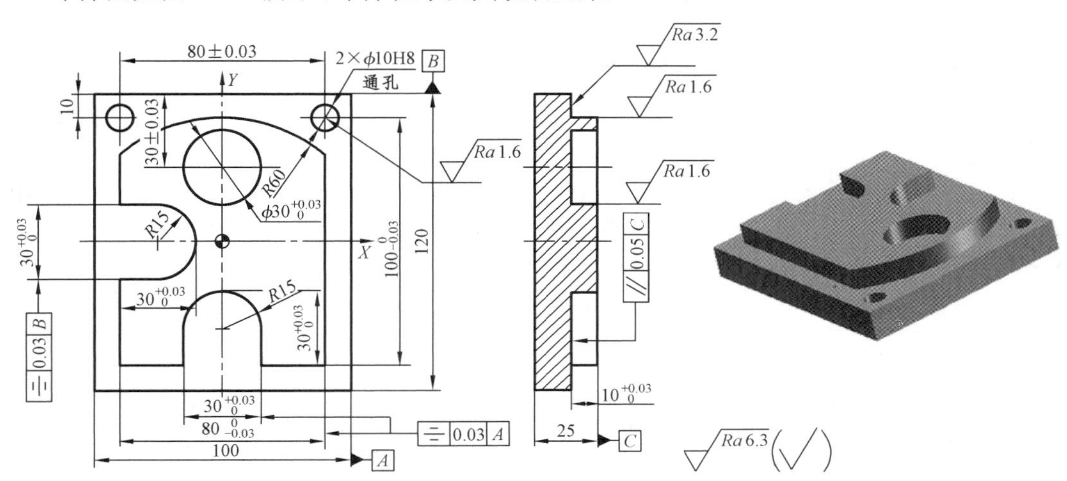

图 2-2-7　零件图

表 2-2-1 零件程序及其说明

程序	说明
O0011	0011 号程序(外轮廓加工)
N10 G90 G94 G21 G40 G54 F100;	设置:绝对坐标编程,每分钟进给,选择毫米制单位输入,刀具半径补偿取消,选择 54 号用户的坐标系,进给速度为 100 mm/min
N20 G91 G28 Z0;	返回 Z 轴参考点
N30 M03 S600;	主轴正转,转速为 600 r/min
N40 G90 G00 X−70.0 Y−80.0;	快速定位至点(X−70.0,Y−80.0)
N50 Z30.0 M08;	快速定位至 Z30.0,打开切削液
N60 G01 Z−10.0;	直线插补至 Z−10.0
N70 G41 G01 X−40.0 Y−80.0 D01;	调用刀具半径补偿,直线插补至点(X−40.0,Y−80.0)
N80 Y−15.0;	直线插补至点(X−40.0,Y−15.0)
N90 X−25.0;	直线插补至点(X−25.0,Y−15.0)
N100 G03 Y15.0 R15.0;	逆时针圆弧插补至点(X−25.0,Y15.0),圆弧半径为 15 mm
N110 G01 X−40.0;	直线插补至点(X−40.0,Y15.0)
N120 Y34.72;	直线插补至点(X−40.0,Y34.72)
N130 G02 X40.0 R60.0;	顺时针圆弧插补至点(X40.0,Y34.72),圆弧半径为 60 mm
N140 G01 Y−50.0;	直线插补至点(X40.0,Y−50.0)
N150 X15.0;	直线插补至点(X15.0,Y−50.0)
N160 Y−35.0;	直线插补至点(X15.0,Y−35.0)
N170 G03 X−15.0 R15.0;	逆时针圆弧插补至点(X−15.0,Y−35.0),圆弧半径为 15 mm
N180 G01 Y−50.0;	直线插补至点(X−15.0,Y−50.0)
N190 X−60.0;	直线插补至点(X−60.0,Y−50.0)
N200 G40 G01 X−80.0 Y−80.0 M09;	刀具半径补偿取消,直线插补至点(X−80.0,Y−80.0),关闭切削液
N210 G91 G28 Z0;	返回 Z 轴参考点
N220 M30;	程序结束并返回程序头
O0012	0012 号程序(钻中心孔)
N10 G90 G94 G21 G40 G54 F50;	设置:绝对坐标编程,每分钟进给,选择毫米制单位输入,刀具半径补偿取消,选择 54 号用户的坐标系,进给速度为 50 mm/min
N20 G91 G28 Z0;	返回 Z 轴参考点
N30 M03 S100;	主轴正转,转速为 100 r/min
N40 G90 G00 Z50.0 M08;	快速定位至 Z50.0,打开切削液
N50 G99 G81 X−40.0 Y50.0 Z−3.0 R−5.0 F50;	钻中心孔循环:刀具快速移动至点(X−40.0,Y50.0),沿 Z 轴快速移动至 R−5.0 参考平面,然后以 50 mm/min 的进给速度沿 Z 轴进给至最终的深度 Z−3.0,最后沿 Z 轴快速返回 R−5.0 参考平面
N60 G98 X40.0 Y50.0;	钻第二个中心孔,最后快速返回 Z50.0 初始平面
N70 X0 Y30.0;	钻第三个中心孔,最后快速返回 Z50.0 初始平面
N80 G80;	钻孔循环取消

续表

程序	说明
N90 G91 G28 Z0;	返回 Z 轴参考点
N100 M30;	程序结束并返回程序头
O0013	0013 号程序(内轮廓加工)
N10 G90 G94 G21 G40 G54 F100;	设置:绝对坐标编程,每分钟进给,选择毫米制单位输入,刀具半径补偿取消,选择 54 号用户的坐标系,进给速度为 100 mm/min
N20 G91 G28 Z0;	返回 Z 轴参考点
N30 M03 S600;	主轴正转,转速为 600 r/min
N40 G90 G00 X0 Y30.0;	快速定位至点(X0,Y30.0)
N50 Z20.0 M08;	快速定位至 Z20.0,打开切削液
N60 G01 Z−10.0 F50;	直线插补至 Z−10.0,进给速度为 50 mm/min
N70 G41 G01 X−5.0 Y30.0 D01 F100;	调用刀具半径补偿,直线插补至点(X−5.0,Y30.0),进给速度为 100 mm/min
N80 G03 X15.0 R5.0;	逆时针圆弧插补至点(X15.0,Y30.0),圆弧半径为 5 mm
N90 G03 I15.0;	(整圆加工)逆时针圆弧插补,从圆弧起点到圆弧中心 X 和 Y 向的矢量分量分别为 15.0 和 0
N100 G40 G01 X0 Y30.0;	刀具半径补偿取消,直线插补至点(X0,Y30.0)
N110 G91 G28 Z0;	返回 Z 轴参考点
N120 M30;	程序结束并返回程序头
O0014	0014 号程序(铰孔)
N10 G90 G94 G21 G40 G54 F50;	设置:绝对坐标编程,每分钟进给,选择毫米制单位输入,刀具半径补偿取消,选择 54 号用户的坐标系,进给速度为 50 mm/min
N20 G91 G28 Z0;	返回 Z 轴参考点
N30 M03 S200;	主轴正转,转速为 200 r/min
N40 G90 G00 Z50.0;	快速定位至 Z50.0,打开切削液
N50 G85 X−40.0 Y50.0 Z−30.0 R−5.0 F50;	镗孔循环:刀具快速移动至点(X−40.0,Y50.0),沿 Z 轴快速移动至 R−5.0 参考平面,然后以 50 mm/min 的进给速度沿 Z 轴进给至最终的深度 Z−30.0,最后沿 Z 轴快速返回 R−5.0 参考平面
N60 X40.0 Y50.0;	铰削第二个孔
N70 G80;	钻孔循环取消
N80 G91 G28 Z0;	返回 Z 轴参考点
N90 M30;	程序结束并返回程序头

2. 自动编程

自动编程软件有很多种,比如 Mastercam、UG、Pro/ENGINEER 等。Mastercam X 是广泛使用的数控自动编程软件。它的主要功能如下:创建几何模型;创建刀具路径;验证切削过程;生成 G 代码。

(1)创建几何模型

可以通过下面两种方式之一创建几何模型:

◎使用 Mastercam X 提供的交互绘图设计功能;

◎在 CAD 软件,比如 UG、Pro/ENGINEER、SolidWorks 中进行设计,然后保存成 Mastercam X 能导入的格式。

本任务中我们将使用 Mastercam X 软件创建几何模型,如图 2-2-8 所示。

图 2-2-8　几何模型

(2)创建刀具路径

◎选择机床类型

点击 机床类型 → 铣床 ,选择机床类型。

◎选择所需的现有刀具

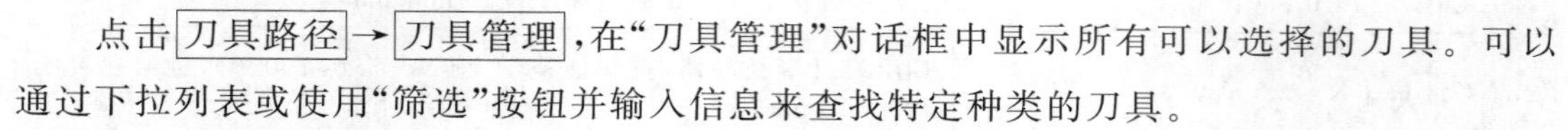

点击 刀具路径 → 刀具管理 ,在"刀具管理"对话框中显示所有可以选择的刀具。可以通过下拉列表或使用"筛选"按钮并输入信息来查找特定种类的刀具。

◎生成刀具路径

•点击主菜单 刀具路径 → 表面粗加工 → 粗加工平行刀具路径 →弹出对话框→选择零件形状为"凸"→点击 ✔ 。

•选择零件被加工表面:单击想要加工的零件表面。确保所有要加工的表面都被选中,然后点击 ✔ 。

•设定表面粗加工平行铣削参数。

(3)验证切削过程

在一个零件被加工之前,需要验证 CAM 模型的零件程序是否正确。验证的目的是:

◎检测刀具路径的几何误差;

◎检测潜在刀具干涉;

◎检测错误的切削条件。

验证开始前,在 刀具路径管理器 中选择一种或多种操作,然后点击验证按钮 ,使用"验证"对话框顶部的控制按钮 ⏮ ▶ ◻ ⏵ ⏭ 进行仿真。验证结果如图 2-2-9 所示。

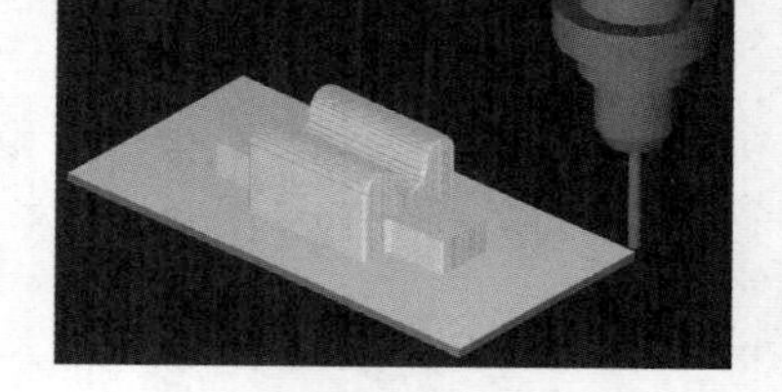

图 2-2-9　验证结果

(4)生成 G 代码

不同的数控机床所使用的 G 代码形式略有不同。将加工数据转换成某一数控机床所需的 G 代码称为后处理。G 代码被存储成不同格式的后处理文件,系统可以使用所选择的任何后处理文件格式。

从 刀具路径标签 的左侧面板中选择图标 G1 →弹出"后处理"对话框→点击 ✔ →输入文件名。生成 G 代码后,就可以查看 G 代码文件,如有需要,还可以进行修改。

Task 3 Machining Center Operation

Speaking Activity

A: Hello, Mr. Li. How are you?

B: Fine, thank you. But I have some problems with machining part on the machining center. Can you do me a favor?

A: Of course. What are these problems?

B: What is the reference point?

A: The reference point which is fixed by the manufacturer is the farthest distance along the positive machine axes.

B: How to return to the reference point?

A: First, ensure the mode selection switch in the "REF" position. Then press the feed axis and direction selection key and home start key until the tool returns to the reference point. When the tool returns to the reference point position, the indicator on the operation panel is on.

B: Why should we establish the workpiece origin offset before machining a part?

A: It's used to compensate for the difference between the programmed workpiece coordinate system and the mechanical coordinate system.

B: Oh, I see. Thank you very much.

A: You are welcome.

Part A Technical Reading

1. The Panel of Machining Center

The panel of machining center usually consists of the control panel and the operation panel.

(1) Control Panel (FANUC Series 0i Mate-MC)

The control panel is shown in Fig. 2-3-1. The functions in the keys in the control panel are seen in the relevant content of Module 1-Task 3.

(2) Operation Panel (FANUC Series 0i Mate-MC)

This task will introduce the machine operation panel of FANUC Series 0i Mate-MC (Fig. 2-3-2).

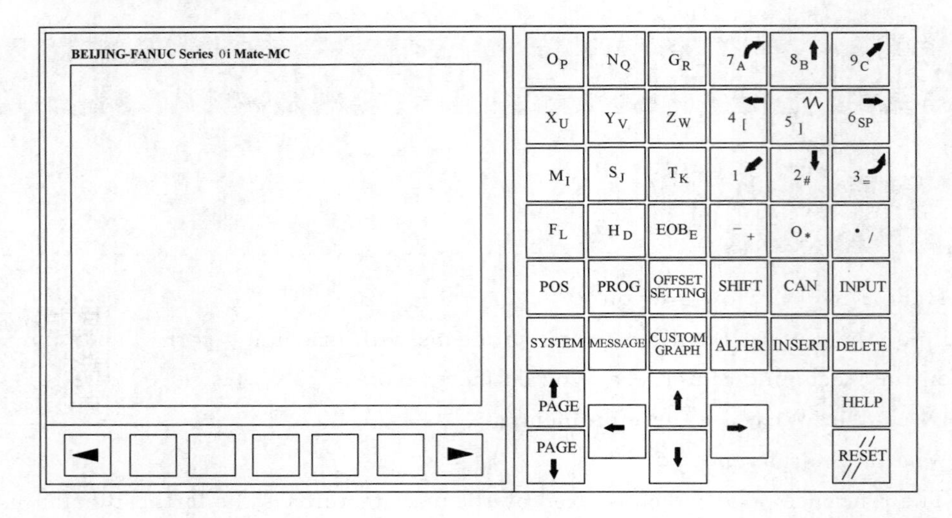

Fig. 2-3-1 Control Panel

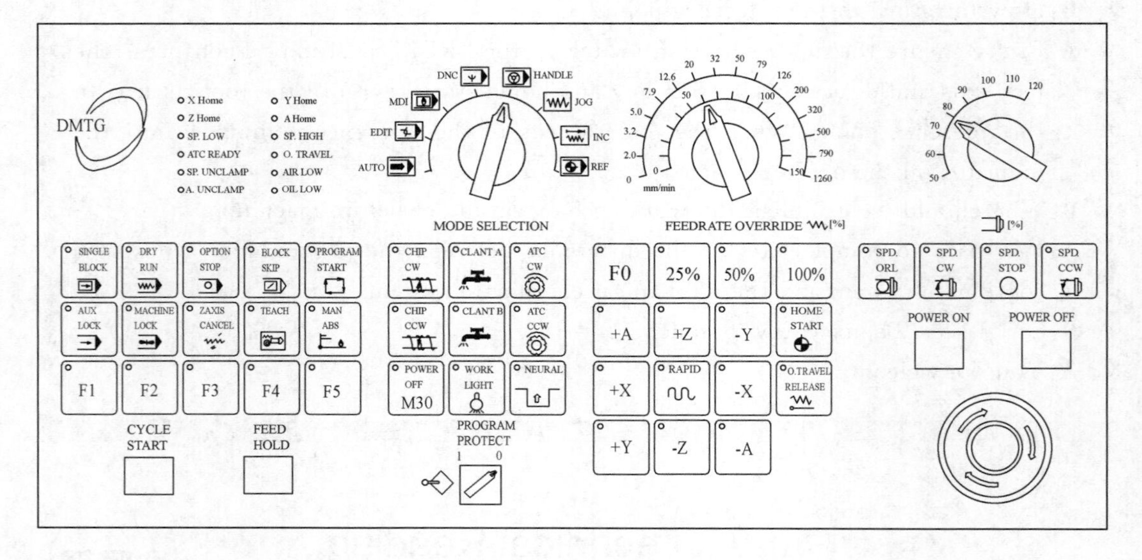

Fig. 2-3-2 Operation Panel

The functions of the keys in the operation panel are shown in Chart 2-3-1.

Chart 2-3-1 The Functions of the Keys in the Operation Panel

Key	Name and Function
AUTO	automatic mode key: make the machine run automatically
EDIT	edit mode key: enter a new program or modify the program
MDI	MDI(manual data input) mode key: enter and execute CNC command by manual
DNC	DNC mode key: direct numerical control

continued

Key	Name and Function
REF	return reference point mode key: return to the reference point position
INC	increment feed mode key: move the tool one step
JOG	jog mode key: manual continuous feed along the X, Y or Z axis
HANDLE	handle mode key: manually move the table with the hand-wheel
mm/min	feed rate knob: adjust the axis feed speed
	spindle rate knob: adjust the spindle speed
SINGLE BLOCK	single block mode key: only one block will be executed
DRY RUN	dry run mode key: check a new program on the machine without any work actually being performed by the tool
OPTION STOP	option stop mode key: press the key M01 to be effective, then press the key M01 again to be invalid
BLOCK SKIP	block skip mode key: the program block signed a tilted bar "/" in the beginning is neglected
PROGRAM START	program start key: make the program restart
MACHINE LOCK	machine lock key: lock the machine and dry run the program
CHIP CW	chip CW key: discharge chips clockwise
CHIP CCW	chip CCW key: discharge chips counterclockwise

continued

Key	Name and Function
ATC CW	ATC CW key: tool storage turns clockwise
ATC CCW	ATC CCW key: tool storage turns counterclockwise
CLANT A CLANT B	cutting fluid control key: turn on or off the cutting fluid
HOME START	home start key: press the key and select X, Y or Z axis, the machine returns to the reference point
SPD. CW	spindle CW key: spindle rotates clockwise
SPD. STOP	spindle stop key: spindle stops
SPD. CCW	spindle CCW key: spindle rotates counterclockwise
CYCLE START	cycle start key: make the program run automatically
FEED HOLD	feed hold key: all feed is interrupted, but the rotation is not affected
	program protection key: in the lock position unauthorized personel cannot modify the program and the parameter, in the unlock position the program and the parameter can be modified
	emergency stop key: make the machine stop under the emergent state
POWER ON	power on key: power on the CNC machine
POWER OFF	power off key: power off the CNC machine

2. An Example of Machining

The part drawing is shown in Fig. 2-3-3. The material of the workpiece is aluminum. We will use the VDF850 machining center for the machining. The flow of machining the part is shown in Fig. 2-3-4.

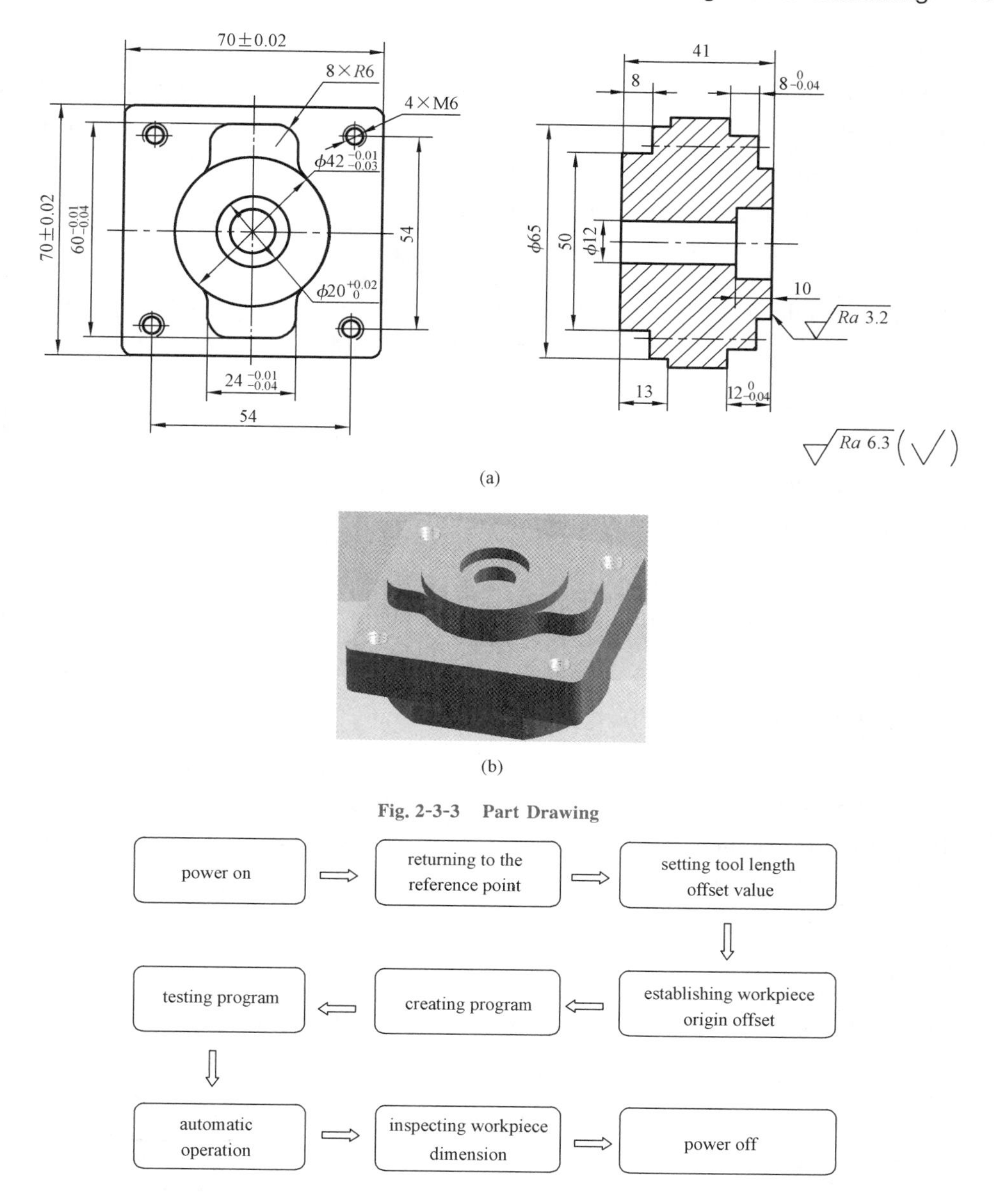

Fig. 2-3-3 Part Drawing

Fig. 2-3-4 Flow of Machining on the Machining Center

(1) Power On

Now that you are familiar with the control panel and operation panel, you may power on the machine. Powering on a FANUC Series 0i Mate-MC machining center will be accomplished in a few steps:

- Switch on the main breaker at the rear of the machine;
- Press the POWER ON key which is located in the lower right corner of the operation panel.

(2) Returning to the Reference Point

The reference point which is fixed by the manufacturer is the farthest distance along the positive machine axes. The machine must find its fixed reference point before it can perform any operation.

The tool can be moved to the reference point by the following ways:

- Turn the mode selection switch 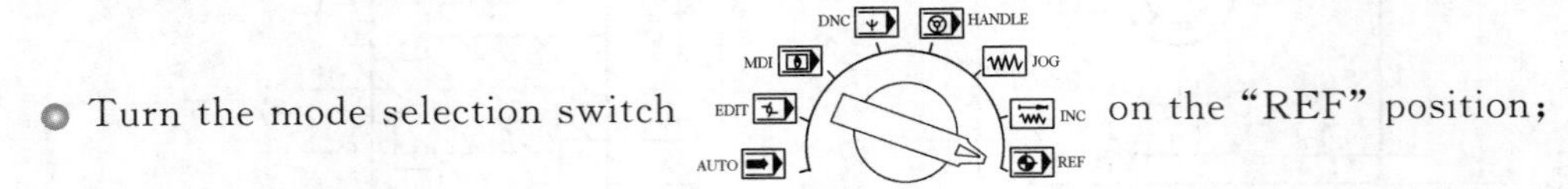

on the "REF" position;
- Press the feed axis and direction selection key and home start key HOME START until the tool returns to the reference point;
- When the tool returns to the reference point, the indicator on the operation panel is on.

For example, if you press the -X and HOME START keys, the X axis returns to the reference point.

(3) Setting Tool Length Offset Value

The procedures for setting tool length offset value are as follows:

- According to the part drawing, mount the desired tool on the spindle;
- Place the height block on the top surface of the workpiece (***Note***: *the thickness of the height block is 50 mm*);
- Use manual operation to move the current tool until it touches the top of the height block (Fig. 2-3-5);
- Press the function key POS, take note of the mechanical coordinate value of Z axis (For example, mechanical coordinate value of Z axis is −421.3);
- Press the offset setting key OFFSET SETTING until the tool length compensation screen is displayed;
- Move the cursor to the compensation number for the current tool and input the correct value −471.3 (***Note***: *the value should be considered for the thickness of the height block in Fig. 2-3-5*).

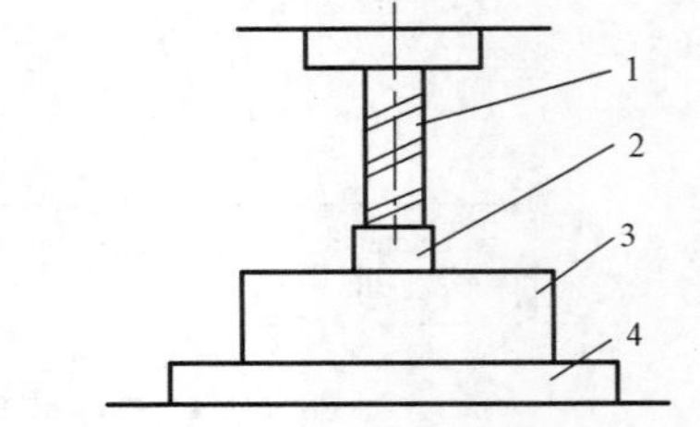

Fig. 2-3-5 Tool Touches the Top of the Height Block

1—tool; 2—height block; 3—workpiece; 4—table

(4) Establishing Workpiece Origin Offset

This function is used to compensate for the difference between the programmed workpiece coordinate system and the mechanical coordinate system. The measured offset for the origin of the workpiece coordinate system can be input on the screen.

◎ Setting *X* Axis Zero

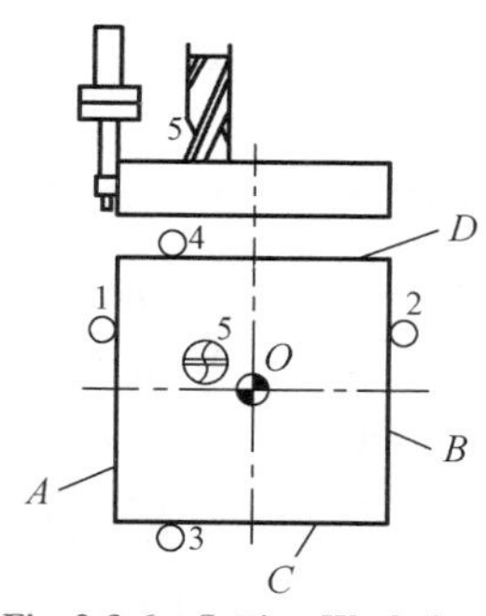

Fig. 2-3-6 Setting Workpiece Origin Cffset

- Mount the workpiece to the table.
- Activate the spindle.
- Move the edge-finder(**Note**: *the radius is* 5 mm) manually until it touches surface *A* of the workpiece(Fig. 2-3-6).
- Retract the tool without changing the *Y* coordinate.
- Set the relative coordinate for *X* axis to zero.
- Use manual operation to move the edge-finder until it touches the surface *B*. Take note of the relative coordinate value of the *X* axis(For example, the relative coordinate value of the *X* axis is 180.80).
- Move the edge-finder to the middle of the workpiece in *X* direction(**Note**: *now the relative coordinate value of the X axis is 90.40*).
- Press [OFFSET SETTING] key until the workpiece origin offset setting screen(Fig. 2-3-7) is displayed. Move the cursor to the "G54 X" register, input "X0", then press the [MEASURE] soft key. This has now set the *X* axis zero for the workpiece.

```
WORKCOORDINATES                               O1234N56789
(G54)

   NO.          DATA           NO.           DATA
   00     X     0.000          02   X        0.000
   (EXT)  Y     0.000          (G55)Y        0.000
          Z     0.000               Z        0.000

   01     X     0.000          03   X        0.000
   (G54)  Y     0.000          (G56)Y        0.000
          Z     0.000               Z        0.000

>Z100.                                   S  0  T0000
MDI **** *** ***                          16:05:59
[NO.SRH] [MEASURE] [         ] [+INPUT]  [INPUT]
```

Fig. 2-3-7 Workpiece Origin Offset Setting Screen

◎ Setting *Y* Axis Zero

The method of setting *Y* axis zero is similar to the *X* axis zero.

(5)Creating Program

Programs can be created using any of the following methods: MDI keyboard; application of transmission software.

This section will describe how to load a program to CNC by using transmission software. The procedures for transmitting the program are as follows:

◎ Make sure the input device is ready for reading;

◎ Turn the mode selection switch 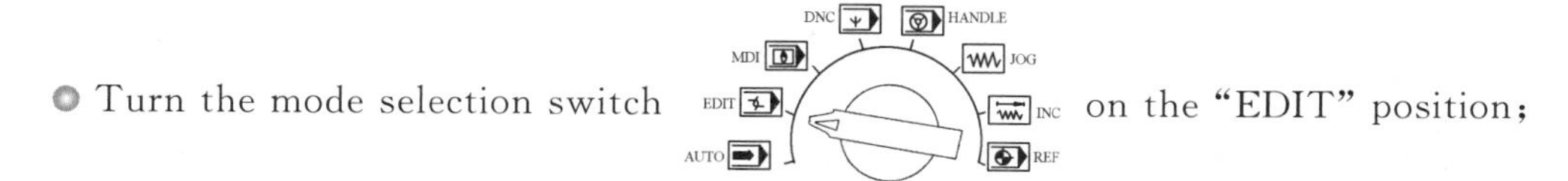

on the "EDIT" position;

● Press the key PROG, then the program content screen appears;

● Press soft key ►;

● After entering address "O", specify a program number to be assigned to the program;

● Press soft key [READ]. The program is input and the program number specified in the previous step is assigned to the program.

(6) Testing Program

Before actual machining, we must check if the machine operates as specified by the created program. The dry run and single block function are commonly used for checking the program.

● Dry Run

When press the dry run mode key DRY RUN on the machine operation panel during automatic operation, the tool moves at the feed rate specified in a parameter. The rapid traverse switch can be used for changing the feed rate.

● Single Block

Press the single block mode key SINGLE BLOCK, then switch to the single block mode. When the cycle start key CYCLE START is pressed in the single block mode, the tool stops after a single block in the program is executed. Check the program in the single block mode by executing the program block by block.

(7) Automatic Operation

● Turn the mode selection switch on the "AUTO" position.

● Press cycle start key CYCLE START, then the program is to be executed.

(8) Inspecting Workpiece Dimension

According to the part drawing, we will use the vernier caliper to inspect the dimension of the workpiece.

(9) Power Off

The most common method is to press the red power off key POWER OFF to cut off the power.

New Words and Phrases

discharge /dɪs'tʃɑːdʒ/	vt. 排出
aluminum /ə'ljuːmɪnəm/	n. 铝
main breaker	主开关
accomplish /ə'kʌmplɪʃ/	vt. 完成,实现,达到
at the rear of ...	在……后面
establish /ɪ'stæblɪʃ/	vt. 建立,设定
secure /sɪ'kjʊə(r)/	vt. 把……弄牢,关紧,使安全
retract /rɪ'trækt/	vt. 缩回,缩进,退回
relative coordinate value	相对坐标值
be similar to ...	与……相似
inspect /ɪn'spekt/	vt. 检查,审查

Part B Practice Activity

Ⅰ. Match A with B.

A	B
参考点	dry run
手轮	emergency stop
空运行	hand-wheel
进给保持	reference point
急停	control panel
创建程序	feed hold
控制面板	create program

Ⅱ. Fill in the blanks according to the text.

1. Reference point which is fixed by the ____________ is the farthest distance along the positive machine axes.

2. The ________________________ and single block functions are commonly used for checking the program.

3. When press the dry run mode key [DRY RUN] on the machine operation panel during ____ ______________, the tool moves at the feed rate specified in a parameter.

4. Check the program in the _____________ mode by executing the program block by block.

5. When press the key [SPD. CCW], the spindle will rotate ____________.

6. The keys [CLANT A] and [CLANT B] are used to ____________________.

7. If you press the [-X] and [HOME START] keys, the *X* axis returns to the ____________________ ____________________.

Ⅲ. Answer the following questions briefly according to the text.

1. Which key is used to adjust the feed rate?

____

2. How do we power on the machine?

____

3. Which methods can be used to create the program?

____

4. Which two functions are commonly used for checking the program?

____

5. Which key is used to discharge chips clockwise?

____

Ⅳ. Translate English into Chinese(Fig. 2-3-8).

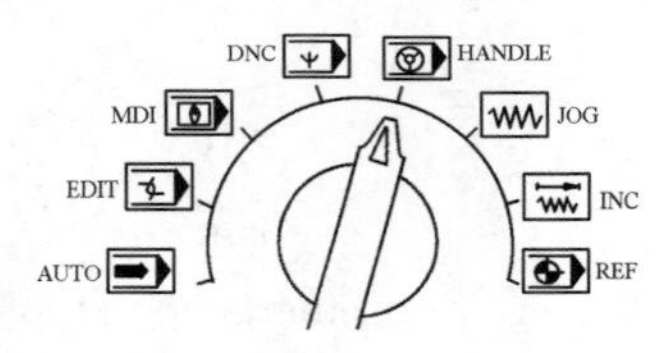

Fig. 2-3-8　Mode Keys

Professional Situation Simulation

Please write a composition in about 150 words according to the following key words.

Key Words: *power on*, *returning to the reference point*, *setting tool length offset value*, *establishing workpiece origin offset*, *creating program*, *testing program*, *automatic operation*, *inspecting workpiece dimension*, *power off*

Work Sheet

....

....

....

....

Part C Broaden Your Horizon

This panel (Fig. 2-3-9) contains all the necessary keys to successfully run the HAAS. However, you must be very familiar with the use of each key before running any component on the machine. If you don't know what the key does, DO NOT PRESS IT. Look through this guide or the HAAS factory user manual or ask one of the staff for assistance.

EDIT	INSERT	ALTER	DELETE	UNDO
MEM	SINGLE BLOCK	DRY RUN	OPT STOP	BLOCK DELETE
MDI DNC	COOLNT	ORIENT SPINDLE	ATC FWD	ATC REV
HANDLE JOG	.0001 .1	.001 1.	.01 10.	.1 100.
ZERO RET	AUTO ALL AXES	ORIGIN	ZERO SINGL AXIS	HOME G28
LIST PROG	SELECT PROG	SEND RS232	RECV RS232	ERASE PROG

Fig. 2-3-9 HAAS Control Panel

Let's begin:

EDIT: Manually edit programs already in memory.

INSERT: Place input data at the location of the cursor or after a highlighted data.

ALTER: Overwrite existing data at cursor location with input value.

DELETE: Delete existing data at cursor location.

UNDO: Undo one previous action.

MEM: Access and/or upload data into active memory.

SINGLE BLOCK: Run program block by block.

DRY RUN: Override feed rates within program.

OPT STOP: Activate optional stop mode(Use M1 command in code).

BLOCK DELETE: Delete entire blocks of code rather than per command.

MDI DNC: (manual data input)enter commands, change tools and activate spindle.

COOLNT: Turn coolant pumps on/off.

ORIENT SPINDLE: Rotate spindle to zero-degree position.

ATC FWD ATC REV: Advance tool changer forward/backward by one tool.

HANDLE JOG: Manually move the table in X, Y and Z.

ZERO RET: Re-zero the machine or send it home location.

AUTO ALL AXES: Re-zero all machine coordinates automatically.

ORIGIN: Set workpiece origin.

ZERO SINGL AXIS: Manually zero selected axis.

HOME G28: Send the machine to part home location.

LIST PROG: Deal entirely with the file system on the machine, from sending and receiving files to erasing them from memory.

SELECT PROG: Upload the selected file into active memory.

SEND RS232: Download selected program to the computer.

RECV RS232: Upload program from computer.

ERASE PROG: Erase selected program.

Translation

任务 3　加工中心零件加工

1. 加工中心面板

加工中心面板通常由控制面板和操作面板组成。

(1)控制面板(FANUC 0i 系列 Mate-MC)

控制面板如图 2-3-1 所示。控制面板上各按键的功能参见模块 1 任务 3 的相关内容。

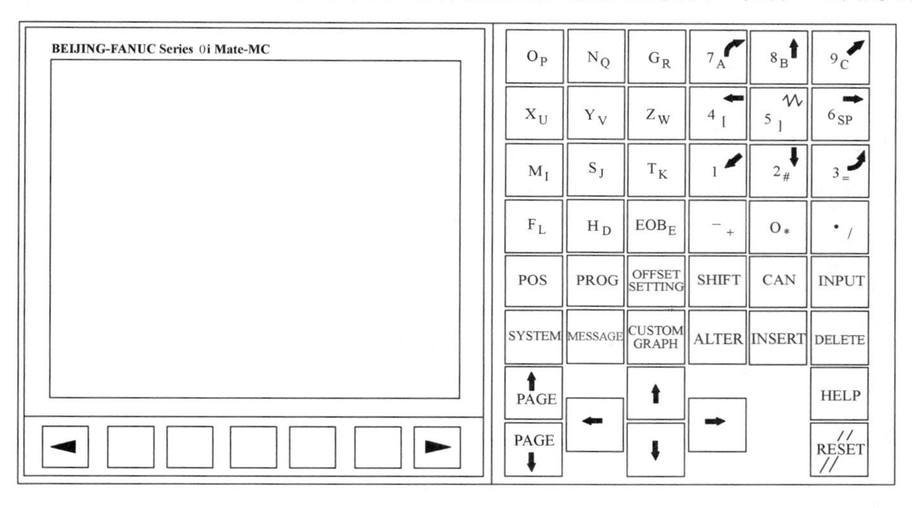

图 2-3-1　控制面板

(2)操作面板(FANUC 0i 系列 Mate-MC)

本任务将介绍 FANUC 0i 系列 Mate-MC 的机床操作面板(图 2-3-2)。

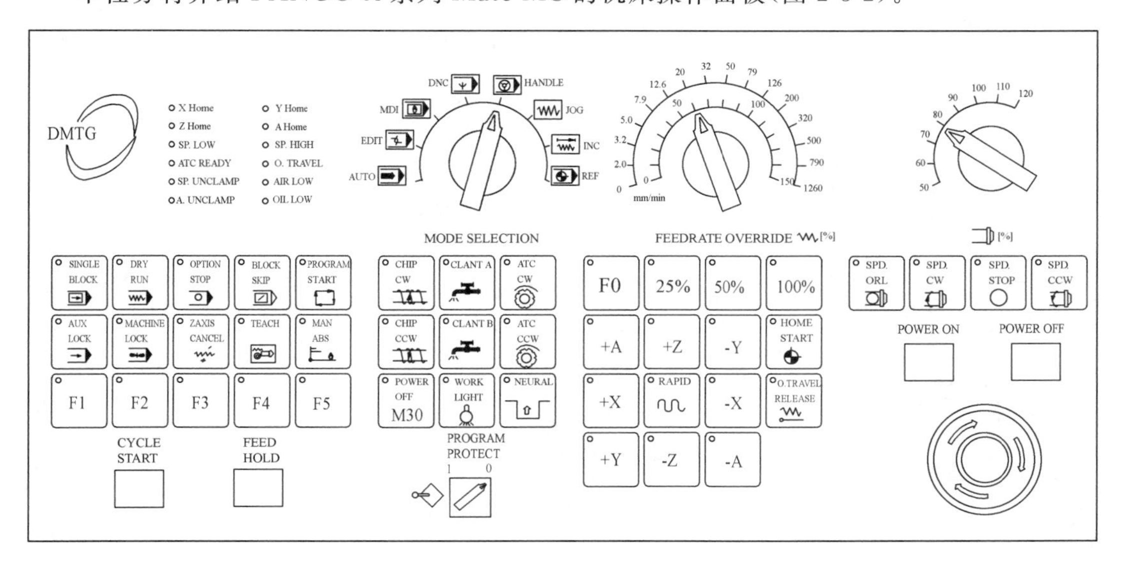

图 2-3-2　操作面板

操作面板上各按键的功能见表 2-3-1。

表 2-3-1　　操作面板上各按键的功能

按键	名称及功能
AUTO	自动模式键:机床自动运行
EDIT	编辑模式键:输入新程序或修改程序
MDI	MDI(手动数据输入)模式键:手动输入和执行数控命令
DNC	DNC 模式键:直接进行数字控制
REF	返回参考点模式键:返回参考点位置
INC	增量进给模式键:使刀具移动一步
JOG	点动模式键:沿 X、Y 或 Z 轴手动连续进给
HANDLE	手轮模式键:使用手轮手动控制工作台的移动
0 2.0 3.2 5.0 7.9 12.6 20 32 50 79 126 200 320 500 790 1260 mm/min; 0 50 100 150	进给速度旋钮:调节坐标轴的进给速度
50 60 70 80 90 100 110 120	主轴速度旋钮:调节主轴转速
SINGLE BLOCK	单段模式键:每次仅执行一个程序段
DRY RUN	空运行模式键:在刀具不切削工件的情况下检查机床上的新程序
OPTION STOP	选择停止模式键:按下 M01 键有效,再次按 M01 键无效
BLOCK SKIP	程序段跳转模式键:前面有“/”的程序段被忽略
PROGRAM START	程序启动键:使程序重新启动
MACHINE LOCK	机床锁住键:锁住机床,空运行程序
CHIP CW	顺时针排屑键:顺时针排屑

续表

按键	名称及功能
CHIP CCW	逆时针排屑键：逆时针排屑
ATC CW	刀库顺时针转动键：刀库按顺时针旋转一个刀位
ATC CCW	刀库逆时针转动键：刀库按逆时针旋转一个刀位
CLANT A CLANT B	切削液控制键：打开或关闭切削液
HOME START	回参考点键：按下此键并选择 X、Y、Z 轴，机床返回参考点
SPD. CW	主轴正转键：主轴顺时针旋转
SPD. STOP	主轴停止键：主轴停止旋转
SPD. CCW	主轴反转键：主轴逆时针旋转
CYCLE START	循环启动键：使程序自动运行
FEED HOLD	进给保持键：所有进给被中断，但不影响主轴旋转
	程序保护键：在锁定状态下未经授权人员许可不能修改程序和参数，在打开状态下才可修改
	紧急停止键：在紧急状态下使机床停止工作
POWER ON	接通电源键：接通数控机床电源
POWER OFF	关闭电源键：关闭数控机床电源

2. 加工实例

零件图如图 2-3-3 所示，工件材料为铝，使用 VDF850 加工中心进行加工。加工零件的流程如图 2-3-4 所示。

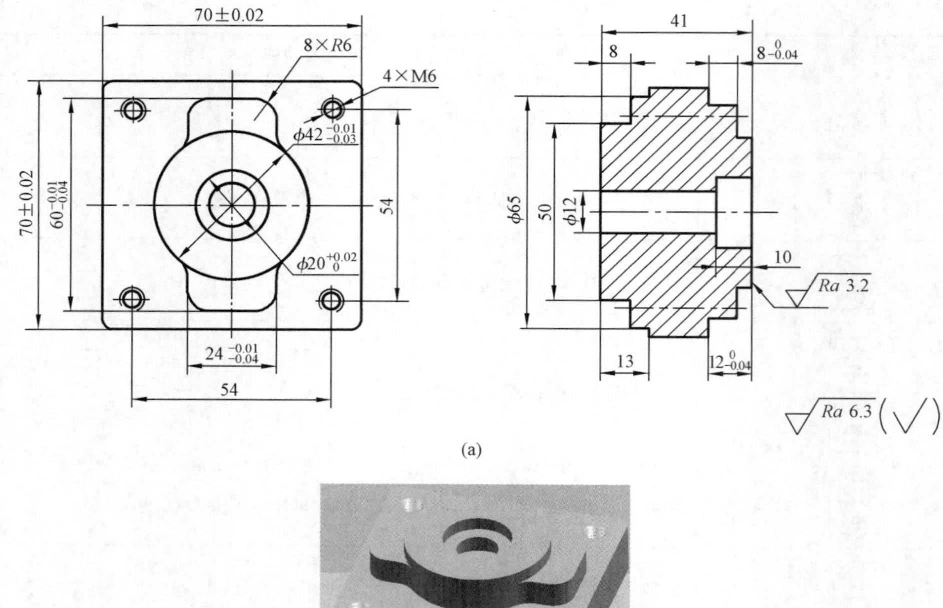

(a)

(b)

图 2-3-3　零件图

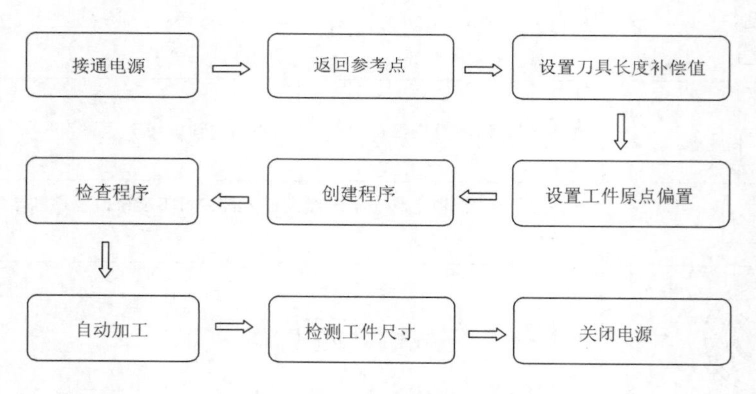

图 2-3-4　加工中心上的加工流程

(1)接通电源

熟悉了控制面板和操作面板后，就可以接通电源了。接通 FANUC 0i 系列 Mate-MC 加工中心需要几个步骤：

- 打开机床后面的主开关；
- 按下位于操作面板右下角的 POWER ON 键。

(2)返回参考点

参考点由机床制造商设定且固定不变，它是机床坐标轴正向的极限点。在机床进行任何操作之前，必须先找到其固定的参考点。

刀具可以通过如下方式返回参考点：

◎将模式选择开关 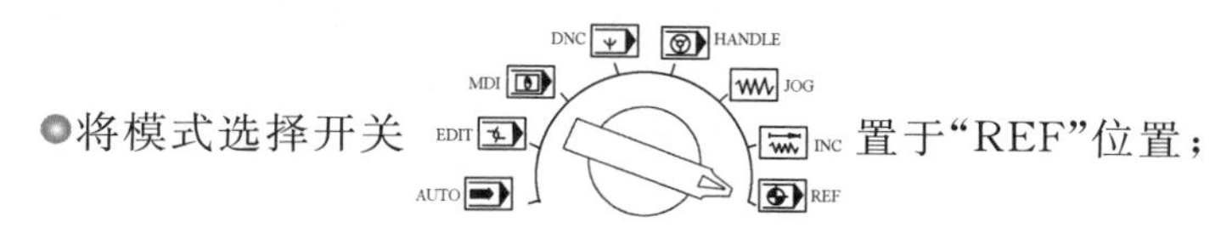

置于“REF”位置；

◎按下进给轴和方向选择键以及回零键 [HOME START]，直到刀具返回参考点；

◎当刀具返回参考点后，操作面板上的指示灯亮。

例如，按下 [-X] 和 [HOME START] 键，X 轴返回参考点。

(3)设置刀具长度补偿值

设置刀具长度补偿值的步骤如下：

◎根据零件图将所需刀具装在主轴上；

◎将高度块放置在工件的上表面上(**注意**：高度块的厚度为 50 mm)；

◎用手动操作移动当前刀具，使其与高度块的上表面接触(图 2-3-5)；

◎按下功能键 [POS]，记录 Z 轴的机床坐标值(例如，Z 轴的机床坐标值为－421.3)；

◎按下参数偏置键 [OFFSET SETTING]，直到显示刀具长度补偿界面；

◎移动光标至当前刀具的补偿号，输入正确的补偿值－471.3(**注意**：该值应考虑图 2-3-5 中高度块的厚度)。

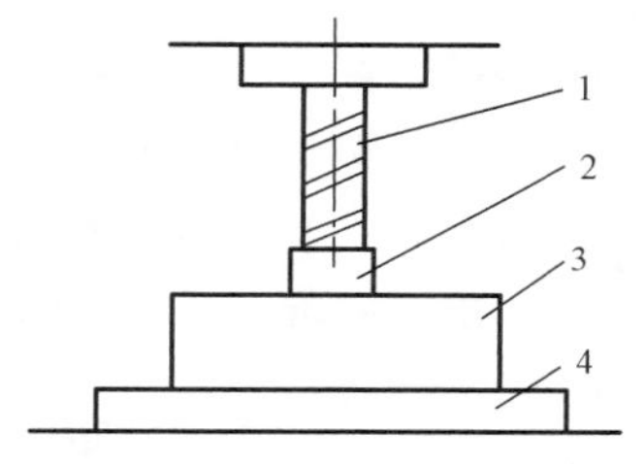

图 2-3-5　刀具与高度块的上表面接触

1—刀具；2—高度块；3—工件；4—工作台

(4)设置工件原点偏置

该功能用来补偿编程的工件坐标系与机床坐标系之间的差值。工件坐标系原点的测量偏置可以通过屏幕输入。

◎设置 X 轴原点

- 将工件安装在工作台上。
- 启动主轴。
- 手动移动寻边器(**注意**：寻边器半径为 5 mm)，直至其与工件的表面 A 接触(图 2-3-6)。
- 使 Y 坐标保持不变，将刀具退回。
- 将 X 轴相对坐标设置为零。
- 手动移动寻边器，直至其与表面 B 接触。记录此时的 X 轴相对坐标值(例如，X 轴的相对坐标值为 180.80)。
- 移动寻边器至工件 X 向的中间位置(注意：此时 X 轴的相对坐标值为 90.40)。

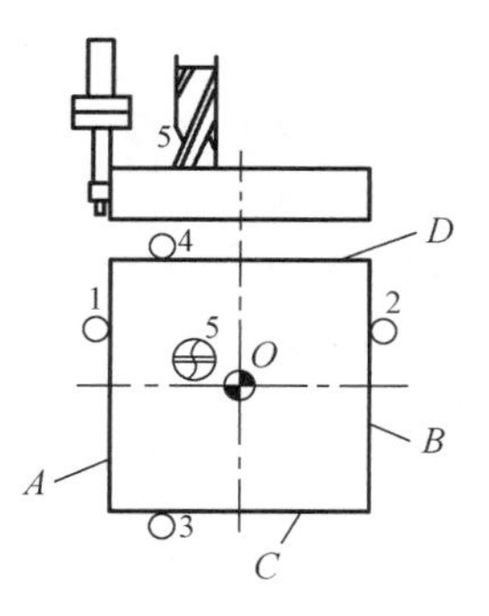

图 2-3-6　设置工件原点偏置

• 按下 OFFSET SETTING 键，直到显示工件原点偏置设置界面(图 2-3-7)。移动光标至“G54 X”处，输入“X0”，然后按下[MEASURE]软键，这样就完成了工件 X 轴零点的设置。

```
WORKCOORDINATES                        O1234N56789
(G54)
     NO.         DATA          NO.          DATA
     00    X     0.000         02  X        0.000
    (EXT)  Y     0.000        (G55) Y       0.000
           Z     0.000              Z       0.000

     01    X     0.000         03  X        0.000
    (G54)  Y     0.000        (G56) Y       0.000
           Z     0.000              Z       0.000

>Z100.                                  S  0  T0000
MDI **** *** ***                        16:05:59
[NO.SRH] [MEASURE] [      ] [+INPUT]  [INPUT]
```

图 2-3-7　工件原点偏置设置界面

◎设置 Y 轴原点

设置 Y 轴原点的方法与设置 X 轴原点的方法相似。

(5)创建程序

可以通过下列方法创建程序：用 MDI 键盘创建程序；用传输软件创建程序。

本部分将介绍如何应用传输软件给数控机床加载程序。传输程序的步骤如下：

◎确认输入设备已经做好读取准备；

◎将模式选择开关 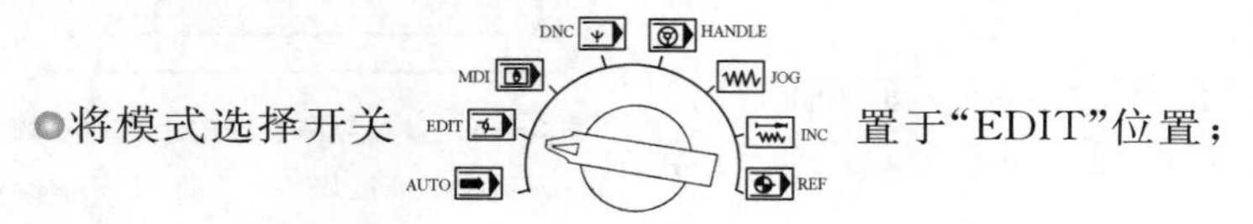

置于“EDIT”位置；

◎按下 PROG 键，显示程序内容界面；

◎按下软键 ► ；

◎输入地址“O”后，指定一个程序号用于分配给程序；

◎按下软键[READ]，程序被输入，上一步中指定的程序号被分配给程序。

(6)检查程序

在实际加工前，必须检查机床是否按照编好的程序进行指定的加工。通常采用空运行和单段功能来检查程序。

◎空运行

在自动运行期间，当按下机床操作面板上的空运行模式键 DRY RUN 之后，刀具按照参数中指定的进给速度运动。快速移动开关可以用来改变进给速度。

◎单段

按下单段模式键 SINGLE BLOCK 切换到单段模式。在单段模式下按下循环启动键 CYCLE START 后，刀具在执行完一段程序后停止。在单段模式下，通过一段一段地执行程序来检查程序。

(7)自动加工

◎将模式选择开关 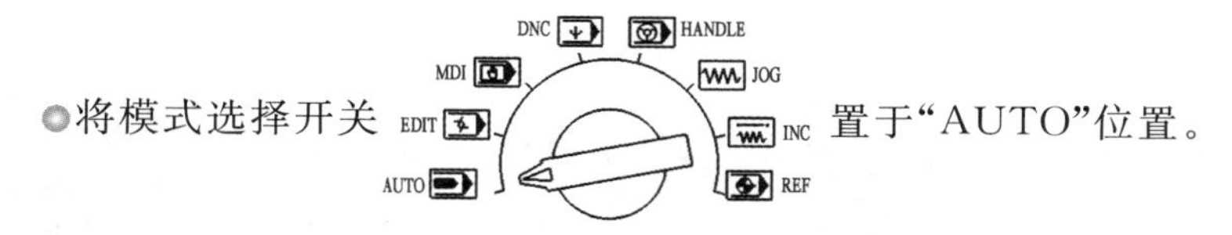

置于“AUTO”位置。

◎按下循环启动键 CYCLE START ，程序被执行。

(8)检测工件尺寸

根据零件图，使用游标卡尺检测工件的尺寸。

(9)关闭电源

最常用的方法是按下红色的关闭电源键 POWER OFF 来切断机床电源。

学习启示

学习执着专注、严谨踏实、一丝不苟、精益求精、追求卓越的工匠精神。

Module 3
CNC Machine Maintenance and Fault Diagnosis Technology

Task 1 Maintenance for CNC Machine

Speaking Activity

A: Hello, I haven't seen you for ages. Where have you been?

B: I've been to a training center for a few days.

A: Oh, really! You went to learn the knowledge of the CNC machine?

B: Yes. I've learned how to operate the CNC machine in the training center.

A: What else?

B: Apart from that, the most important is that I've learned some knowledge about warning marks and maintenance for CNC machine.

A: What are the potential hazards in a metal cutting operation?

B: Main potential hazards include: rotating components, movable components, errors in the program and an error in setting or changing the offset value.

A: How to avoid errors in the program?

B: You must use the machine lock, dry run and single block function to confirm each line in the program before executing it.

A: Do we need to stop the machine when removing the long curled chips?

B: Yes, we must stop the machine and use chip hooks to break the long curled chips.

A: How often do we replace the coolant?

B: Every eight hours.

A: Today I've got a lot. Thank you!

B: It is my pleasure to talk with you.

Part A Technical Reading

1. Safety Notes

Safety is always a major concern in a metal-cutting operation. CNC equipment is automatic and very fast and consequently it is a source of hazards. The hazards have to be located and the operator must be on guard of them in order to prevent injuries and damages to the equipment. Main potential hazards include: rotating components, such as the spindle, the tool in the spindle, chuck, workpiece in the chuck and the turret with the tools and rotating clamping devices; movable components, such as the machining center table, lathe slides, tailstock center and turret; errors in the program, such as improper use of the G00 code in conjunction with the wrong coordinate value, which can generate an unexpected rapid motion; an error in setting or changing the offset value, which can result in a collision of the tool with the workpiece or the machine; a hazardous action of the machine caused by changing a proven program randomly. The preventive actions should be followed to minimize or avoid hazards:

(1) Keep all of the original covers on the machine supplied by the machine manufacturer.

(2) Wear safety glasses, gloves, proper clothing and shoes.

(3) Do not attempt to run the machine until you are familiar with the proper operation procedures.

(4) Before running the program, make sure that the workpiece is clamped properly.

(5) When proving a program, follow these procedures:

- Run the program in the machine lock function to check the program for errors in syntax and tool path;
- Use the dry run function to check the program;
- Use a single-block execution to confirm each line in the program before executing it;
- While the tool is cutting, slow down the feed rate using the rapid override switch to prevent excessive cutting conditions.

(6) Do not handle chips by hand and do not use chip hooks to break long curled chips. Choose different cutting conditions for better chip control. Stop the machine if you need to properly clean chips.

(7) If there is any doubt that the insert will break under the programmed cutting conditions, choose a thicker insert or reduce feed or depth of cut.

(8) Keep tool overhang as short as possible, since it can be a source of vibration that can break the insert.

(9) When supporting a large workpiece by the tailstock center, make sure that the

center-hole is large enough to adequately support and hold the workpiece.

(10) Stop the machine when changing the tools, indexing inserts or removing chips.

(11) Replace dull or broken tools or inserts in time.

(12) Write a list of offsets for active tools and clear the offsets for tools removed from the machine.

(13) Do not make changes in the program if you are not permitted by your supervisor.

(14) If you have any safety-related concerns, notify your instructor or supervisor immediately.

2. Daily Maintenance

Daily maintenance is one of the important maintenance ways. If you often maintain your machine and read the maintenance manual, you will be able to deal with many minor issues in your work. These required specifications must be followed in order to keep your machine in good working condition:

(1) Daily Maintenance

- Fill up the coolant every eight hours.
- Check lubrication tank level of guide way.
- Clean chips from way covers and bottom pan.
- Clean chips from tool changer.
- Wipe spindle taper with a clean cloth rag and apply lubricating oil.

(2) Weekly Maintenance

- Check for normal working of auto drain on filter regulator.
- Disclose the tank cover and remove any sediment inside the tank.
- Check the pressure of the air gauge/regulator.

(3) Monthly Maintenance

- Check oil level in the gear box.
- Clean pads on bottom of pallets.
- Inspect way covers for proper operation and lubricate with lubricating oil, if necessary.

(4) Half-year Maintenance

- Replace coolant and thoroughly clean the coolant tank.
- Check all hoses and lubrication lines for cracking.

(5) Annual Maintenance

- Replace the gear box oil.
- Check the oil filter and clean out residue at the bottom of filter.

3. Warning Marks

Warning marks are usually placed on CNC machines in locations where hazards exist. We must understand the explanation of each warning mark. Common warning marks are shown in Chart 3-1-1.

Chart 3-1-1 **Common Warning Marks**

Warning mark	Explanation	Warning mark	Explanation
	high pressure risk		pay attention to dust
	pay attention to safety		beware of high temperature
	be careful of noise		warning forklift
	warning mechanical injury		warning overhead load
	caution falling objects		warning splinter
	no switching, it is working		no climbing, high pressure danger
	prohibit the operation, it is working		no approaching
	must wear a safety cap		must wear protective glasses

4. Labor Protection

Labor protection is a general term for the legislation, organization and technical measures taken by the state and the unit to protect the safety and health of workers in the process of labor production. The purpose is to create a safe labor protection for workers, healthy and comfortable working conditions, eliminate and prevent the possibility of casualties in the production process of labor, occupation disease and acute poisoning. Commonly used protective measures are as follows:

(1) Eye Protection

Using eye protection in the machine shop is the most important safety rule of all. Metal chips or shavings can fly at a high speed and cause serious eye injury.

(2) Hazardous Noise Protection

Noise hazards are very common in the machine shop. High intensity noise can cause permanent loss of hearing. Although noise hazards cannot always be eliminated, hearing loss is avoidable with ear muffs and ear plugs.

(3) Foot Protection

The floor in a machine shop is often covered with razor-sharp metal chips, and heavy stock may drop onto the feet. Therefore, safety shoes or solid leather shoes must be worn all the time. These have steel plates located over the toes and are designed to resist impact. Some safety shoes also have an instep guard.

Electrical workers must wear qualified insulation shoes. Insulation shoes as safety auxiliary tool engaged in electrical work, can effectively reduce the chance of electric shock.

New Words and Phrases

hazard /ˈhæzəd/	*n.* 危险，危害
consequently /ˈkɒnsɪkwəntli/	*adv.* 结果，因此，必然地
collision /kəˈlɪʒn/	*n.* 碰撞，相撞
randomly /ˈrændəmli/	*adv.* 任意地，随便地，胡乱地
execution /ˌeksɪˈkjuːʃn/	*n.* 实行，执行，履行，完成
curled /kɜːld/	*adj.* 卷曲的
adequately /ˈædɪkwətli/	*adv.* 适当地，足够地，充分地
maintenance /ˈmeɪntənəns/	*n.* 维护，保养
top off	装满，平仓
slide way	导轨

hose /həʊz/	*n.* 橡皮软管
rag /ræg/	*n.* 破布，碎布
drain /dreɪn/	*n.* 排水，下水道，排水系统
sediment /ˈsedɪmənt/	*n.* 沉淀物，沉积物，沉渣
residue /ˈrezɪdjuː/	*n.* 残渣
forklift /ˈfɔːklɪft/	*n.* 叉车
splinter /ˈsplɪntə/	*n.* 碎片，尖片
casualties /ˈkæʒjuəltiz/	*n.*（casualty 的复数）伤亡者，牺牲品，受害者
razor-sharp	*adj.* 锋利的，犀利的
instep /ˈɪnstep/	*n.* 脚背，脚背形的东西

Part B Practice Activity

Ⅰ. Match A with B.

A	B
职业病	daily maintenance
日常保养	occupation disease
劳动保护	acute poisoning
急性中毒	labor protection
危险源	top off coolant
装满冷却液	source of hazards

Ⅱ. Answer the following questions briefly according to the text.

1. What are the main potential hazards of the CNC equipment?

____

____

2. What will you do first if you need to properly clean chips?

____

3. Why does the operator keep tool overhang as short as possible?

____

4. What will you do under the bad chip control condition?

____

5. Which function will you choose to confirm each line in the program before executing it?

____

Ⅲ. Match the pictures with the explanations.

A. Warning splinter B. No switching, it is working

C. Must wear protective glasses D. Warning forklift

1

2

3

4

Ⅳ. Mark the following statements with T(true) or F(false).

(   )1. Warning marks are usually placed on CNC machines in locations where hazards exist.

(   )2. Metal chips can fly at a slow speed and cannot cause serious eye injury.

(   )3. Some safety shoes also have functions for instep guarding.

(   )4. Low intensity noise can cause permanent loss of hearing.

(   )5. Commonly used protective measure is only hazardous noise protection.

Professional Situation Simulation

Please write a composition in about 150 words according to the following key words.

Key Words: *safety*, *hazards*, *damages*, *safety decals*, *daily maintenance*, *check*

Work Sheet

....

....

....

....

....

....

....

....

....

Part C Broaden Your Horizon

All maintenance shops and work areas should be marked with the correct colors to identify hazards, exits, safe walkways and first-aid stations. It is acceptable to use materials other than paint, such as decals and tapes, in the appropriate, similar colors. Listed below are the main colors authorized using in maintenance shops.

Red color markings should be used to identify the following equipments or locations:

- Fire alarm boxes;
- Fire blanket boxes;
- Fire pumps;
- Emergency stop buttons for electrical machinery;
- Emergency stop bars on hazardous machines;
- Caution signs.

Green color markings normally on a white color background should be used for the following equipments or locations:

- First-aid equipments and first-aid dispensaries;
- Safety starting buttons on machinery.

Orange markings are used to designate dangerous components of machines or energized equipment, including electrical conduits, which may cut, crush, shock or injure people.

Translation

模块3 数控机床的维护与故障诊断技术

任务1 数控机床的维护

1. 安全操作规程

金属切削操作中的安全性一直受到关注。由于数控设备自动化程度高并且速度快，所以它是一个危险源。为了防止人员伤害和对设备的损坏，必须找出危险的根源，且操作人员必须提高警惕。主要潜在的危险包括：旋转部件，如主轴、装夹在主轴上的刀具、卡盘、卡盘上的工件、装有刀具的刀架以及旋转夹具装置；运动部件，如加工中心工作台、车床拖板、尾座顶尖和刀架；程序错误，如由G00指令的不正确使用引起的坐标值错误，这会产生意想不到的快速移动；设置或改变偏移值时出错，这会导致刀具与工件或刀具与机床之间的碰撞；随意地更改已验证的程序，会使机床产生危险动作。为了减少或避免危险，应尽量采取以下保护措施：

(1)保护好机床制造商提供的所有机床原有防护罩。

(2)戴上安全眼镜、手套,穿上合适的衣服和鞋。

(3)在不熟悉机床的正确操作步骤的情况下,不要开动机床。

(4)运行程序之前,确认工件已被正确夹紧。

(5)验证程序时应遵循下列步骤:

- 启用机床锁定功能中的程序,以检查程序中的语法错误和刀具路径;
- 使用空运行功能检查程序;
- 在执行程序前,采用单段执行来确定程序中的每一行;
- 刀具切削时,用进给倍率开关来降低进给速度,防止超负荷切削。

(6)禁止用手处理切屑以及用切屑钩子弄断长而卷曲的切屑。选择不同的切削条件以便更好地控制切屑。如果需要彻底清除切屑,则应当关闭机床。

(7)如果怀疑刀片在编程的切削状态下有可能折断,则可选择一个更厚的刀片或减小进给速度或切削深度。

(8)刀具伸出尽可能短些,因为它可能成为导致刀片折断的振动源。

(9)当用尾座顶尖支撑大工件时,应确保中心孔足够大,足以支撑并夹住工件。

(10)更换刀具、替换刀片或清理切屑时一定要关闭机床。

(11)及时替换已磨损或损坏的刀具或刀片。

(12)写下激活刀具的偏移量清单,对于从机床上卸下的刀具,应清除刀具偏移量。

(13)在未得到主管许可的情况下,不得擅自更改程序。

(14)如果你有任何与安全有关的担忧,马上通知你的技术指导或主管。

2. 日常保养

日常保养是一种重要的保养方法。如果你经常保养你的机床并阅读保养手册,你就能够处理在工作中出现的很多小问题。为了使机床正常工作,必须遵守这些必要的规范:

(1)每日保养

- 每 8 小时加满冷却液。
- 检查导轨润滑油箱的液位。
- 清理导轨防护罩和地板上的铁屑。
- 清理换刀装置上的铁屑。
- 用干净的碎布清洁主轴锥孔,并涂上润滑油。

(2)每周保养

- 检查过滤器减压阀是否正常工作。
- 卸下冷却箱盖子,清理其中的沉淀物。
- 检查气压表/调节器的压力。

(3)每月保养

- 检查齿轮箱中的油位。
- 清理托盘底部的衬垫。
- 检查导轨防护罩是否正常运行,必要时用润滑油润滑。

(4)每半年保养

- 更换冷却液,彻底清洗冷却箱。
- 检查所有软管和润滑油路是否破裂。

(5)每一年保养

- 更换齿轮箱润滑油。
- 检查润滑油过滤器,清除过滤器底部的残余物。

3. 警示标志

数控机床上存在危险的地方通常设有警示标志。我们必须理解每个警示标志的含义。

常用警示标志见表 3-1-1。

表 3-1-1 常用警示标志

警示标志	含义	警示标志	含义
	高压危险		注意防尘
	注意安全		当心高温
	当心噪声		当心叉车
	当心机械伤人		当心吊物
	当心落物		当心扎脚
	禁止合闸，有人工作		禁止攀登，高压危险
	禁止操作，有人工作		禁止靠近
	必须戴安全帽		必须戴防护眼镜

4. 劳动保护

劳动保护是国家和单位为保护劳动者在劳动生产过程中的安全和健康所采取的立法、组织和技术措施的总称。劳动保护的目的是为劳动者创造安全、卫生、舒适的工作条件，消除和预防生产过程中可能发生的伤亡、职业病和急性中毒。常用防护措施如下：

(1)眼部防护

在生产车间里，采取眼部保护措施是所有的安全规则中最为重要的一项。金属切屑或刨花能以很快的速度飞出，导致严重的眼部损伤。

(2)危害性噪声防护

在生产车间里,噪声危害是非常普遍的。高频噪声能引起永久性听力损伤。虽然噪声危害不可能被消除,但是通过戴减噪耳套、防噪声耳塞,都可避免听力损伤。

(3)足部防护

在生产车间里,地面经常被锋利的金属切屑覆盖,并且重型的储备物也可能砸到脚部。因此,安保鞋或坚固的皮鞋应该一直穿着。这些鞋在脚尖的部位都有一块抗压钢板。有些安保鞋还有脚背部防护。

电工作业时必须要穿戴合格的绝缘鞋。绝缘鞋作为从事电气工作的安全辅助用具,可以有效地降低触电的概率。

Task 2 CNC Machine Fault Diagnosis Technology

Speaking Activity

A: Tom, how are you getting on with your work recently?

B: Fine, and you? How about your work?

A: Not very good. I am very puzzled about the CNC machine fault diagnosis. Can you say something about it?

B: Of course. Fault diagnosis methods include direct-vision, alarm function, replacing parts, principle analysis and so on.

A: Good. I am especially interested in the alarm function. Can you explain it in detail?

B: Yes. Each machine has an alarm function. The alarm function of CNC machine can help people decide whether something is wrong with the machine. As long as we understand the content of the alarm message, the fault can be ruled out quickly and accurately.

A: What methods do you use to rule out the fault related to PLC technology?

B: We can rule out the fault related to PLC technology according to working principle of control object, I/O status of PLC and the alarm message.

A: Oh, I see. Thank you very much.

B: You are welcome.

Part A Technical Reading

1. Fault Diagnosis Method

When a failure of CNC machine occurs, it is important to examine the real cause and take appropriate action. To do this, we usually adopt the following methods:

(1) Direct-vision

- Asking

Ask the operator about the operation state when the trouble occurs and the whole course of trouble. Find out whether it is suddenly or gradually happening and also ask whether the trouble happened before and so on.

- Looking

Determine the location where it results in troubles by observing the rotation speed of the main drive spindle, the color of the cooling liquid, damage to machined parts, drive axes and ball screw.

- Listening

Determine what is wrong with gear coupler, ball screw and nuts by listening to the location where unusual sound emerges when the machine is running.

- Touching

Touch the machine and tell whether the machine is too hot or vibrating.

- Smelling

The smell of something burning can help decide whether the machine has been affected by violent friction or whether the electrical components have been burnt up.

(2) Alarm Function

Each CNC machine has an alarm function. The alarm function of CNC machine can help people decide whether something is wrong with the machine. Some alarm messages of CNC machine are shown in Chart 3-2-1.

Chart 3-2-1 Alarm Messages of CNC Machine

number	message	fault content	remedy
010	improper G code	an unusable G code or G code corresponding to the function not provided is specified	modify the program
073	program number already in use	the commanded program number has already been used	change the program number or delete unnecessary programs and execute program registration again
100	parameter write disenable	on the PARAMETER(SETTING) screen, PWE(parameter writing enable) is set to 1	set it to 0, then reset the system

continued

number	message	fault content	remedy
150	illegal tool group number	tool group number exceeds the maximum allowable value	modify the program
180	communication error (remote buffer)	remote buffer connection alarm has generated	confirm the number of cables, parameters and I/O device

(3)Functional Program Test

Make a function test program with GMST function of CNC system to determine correctness and precision of the movement.

(4)Replacing Parts

If there is a suspicion that some part has trouble, replace it with a spare part to decide whether the machine has some troubles before the replacement of the part.

(5)Checking Parameters

Both the loss of the system parameters and incorrect settings may cause changes in the function of the machine or other troubles. Therefore, once the machine has some troubles, check and correct the system parameters timely.

(6)Principle Analysis

Use the working principle to analyze the trouble and find out the source of trouble with the CNC machine.

2. Fault Diagnosis and Treatment of CNC Machine Based on PLC Technology

PLC is an important component of modern CNC system. It is a bridge between CNC system and machine. In order to rule out the fault related to PLC timely, we need to use several ways to analyze the fault synthetically and narrow the scope of failure gradually and rule out the fault finally.

(1)Fault Diagnosis According to the Working Principle of Control Object

PLC procedures of a CNC machine are designed in accordance with the working principle of control object. It is a very effective method to do fault diagnosis through analyzing the working principle of control objects and the I/O status of PLC. The example is shown as follows:

◎ **Example 1**: To a CNC lathe equipped with FANUC OTC system, when tramping on the footswitch, the workpiece cannot be chucked.

• **Fault analysis**: In accordance with machine working principle, the first time when the footswitch is tramped, the workpiece should be chucked, and when the footswitch is tramped second time, the workpiece should be released. Let the footswitch connect with PMC display screen and find that the input $X2.2$ is zero all the time after the footswitch is tramped. So we can estimate that there is something wrong with the footswitch.

• **Fault treatment**: Check footswitch and replace it with a new one.

(2) Fault Diagnosis According to I/O Status of PLC

In a CNC machine, transmission of input/output signal usually achieves through the I/O interface of PLC. So many failures will be reflected in I/O interface of PLC. It is convenient to do fault diagnosis according to such characteristics. If there is no hardware failure, we can directly check I/O interfaces of PLC to identify the cause of the fault without checking ladder graph and circuit diagram. The example is shown as follows:

Example 2: There is a fault in a CNC machine that the protective door cannot be closed and the machine cannot do automatic processing.

- **Fault analysis**: The protective door is opened by the gas cylinder and it is closed by the electromagnetic valve YV2. 0 controlled by PLC output Q2. 0. Check the status of Q2. 0 and find it "1", but electromagnetic valve YV2. 0 is not in power. So we can say that the intermediate relay KA2. 0 is damaged because PLC output Q2. 0 controls electromagnetic valve YV2. 0 by it.
- **Fault treatment**: Replace a new intermediate relay. Then the fault is ruled out after replacement of relay.

(3) Fault Diagnosis According to the Alarm Message

Modern CNC system is rich in self-diagnosis function and the fault alarm message is displayed on the CRT. As long as we understand the content of alarm message, the CNC equipment fault can be ruled out quickly and accurately. The example is shown as follows:

Example 3: A CNC machining center equipped with SINUMERIK 820 CNC system gives an alarm of 7035 and relative message in the report is that the graduator of sub-table does not set down.

- **Fault analysis**: In SINUMERIK 810/820 CNC system, the alarm of "7" prefix is set by PLC operational information or machine plant. Approach of dealing with the fault is to call up PLC input/output status and compare it with the copy list according to the information of the fault. Whether the graduator of sub-table set down or not is detected by the switch SQ25, SQ28 under the table. SQ28 detects whether the graduator rotates in place and it is related to PLC inputs I10. 6. SQ25 detects whether the graduator set down in place and it is related to PLC inputs I10. 0. The graduator setting down is completed by the electromagnetic valve YV06 driven by relay KA32 through the output interface Q4. 7. Observe from the PLC STATUS: if I10. 6 is "1", it shows that the graduator rotates in place, and if I10. 0 is "0", it shows that the graduator does not set down. Then observe Q4. 7, if it is "0", KA32 relay shall not be electrified and electromagnetic valve YV06 does not work, so it alarms because the graduator does not set down.
- **Fault treatment**: Move electromagnetic valve YV06 manually and observe whether the graduator sets down to distinguish the fault whether in output circuit or in the internal PLC.

3. Fault Diagnosis Device—SKF Equipment Condition Monitor

(1) Overview

Now both novice users and experts can easily, quickly and accurately check the condition of rotating equipment by using the SKF Equipment Condition Monitor.

The SKF Equipment Condition Monitor provides an overall vibration velocity reading that measures vibration signals from the machine and automatically compares them to pre-programmed International Standardization Organization (ISO) guidelines. An "Alert" or "Danger" alarm displays when measurements exceed those guidelines. Simultaneously, an "Enveloped Acceleration" measurement is taken and compared with established bearing vibration guidelines to verify conformity or indicate potential bearing damage. The SKF Equipment Condition Monitor also measures temperature using an infrared sensor to indicate uncharacteristic heat.

(2) Features

- Quick and easy to set up and use.
- Lightweight, compact and ergonomically designed.
- Alert and danger prompts provide increased diagnostic confidence.
- Measuring vibration velocity, enveloped acceleration and temperature simultaneously to save time.
- Efficient, economical and environmentally friendly, the rechargeable SKF Equipment Condition Monitor operates 10 hours on a single charge.
- An optional external sensor can be used for hard-to-reach locations.

New Words and Phrases

diagnosis /ˌdaɪəgˈnəʊsɪs/	*n.* 诊断,诊断结论,判断,结论
coupler /ˈkʌplə/	*n.* 联轴器
alarm /əˈlɑːm/	*n.* 惊恐,警报
consistent /kənˈsɪstənt/	*adj.* 一致的,一贯的,连续的
remedy /ˈremədi/	*n.* 治疗法,补救办法,纠正办法
prefix /ˈpriːfɪks/	*n.* 前缀
distinguish /dɪˈstɪŋgwɪʃ/	*vi.* 区分,辨别,分清;*vt.* 区分,辨别,分清
novice user	初学者
alert /əˈlɜːt/	*n.* 警戒,警报,警惕
enveloped acceleration	加速度包络
infrared sensor	红外传感器

Part B Practice Activity

Ⅰ. March A with B.

A	B
故障诊断	novice user
报警功能	control object
系统参数	electromagnetic valve
原理分析	system parameter
控制对象	fault diagnosis
I/O 接口	I/O interface
电磁阀	principle analysis
初学者	alarm function
红外传感器	infrared sensor

Ⅱ. Mark the following statements with T(true) or F(false).

(　　) 1. Don't ask the operator about the operation state when the trouble occurred and the whole course of trouble.

(　　) 2. Some machines have an alarm function but others do not.

(　　) 3. Don't touch the machine to tell whether the machine is too hot or vibrating.

(　　) 4. Determine the location which results in troubles by observing the rotating speed of the main drive spindle.

(　　) 5. In order to rule out the fault related to PLC timely, we need to use several ways to analyze the fault synthetically.

(　　) 6. SKF Equipment Condition Monitor cannot measure temperature.

Ⅲ. Answer the following questions briefly according to the text.

1. How many methods can be used to find out the fault of CNC machines? What are they?

____

____

2. What is the meaning of alarm message number 073? What method can you take to get remedies?

____

____

3. How many methods do you use to rule out the fault related to PLC technology?

____

4. What's the function of SKF Equipment Condition Monitor?

____

Ⅳ. Translate the following sentences into Chinese.

1. If there is a suspicion that some part has trouble, replace it with a spare part to decide whether the machine has some troubles before the replacement of the part.

____

____

2. In CNC machines, transmission of input/output signal usually achieves through the PLC I/O interface. So many failures will be reflected in I/O interface of PLC.

____

____

3. PLC procedures of CNC machine are designed in accordance with the principle of control object.

____

4. Simultaneously an "Enveloped Acceleration" measurement is taken and compared to established bearing vibration guidelines to verify conformity or indicate potential bearing damage.

____

Professional Situation Simulation

Please write a composition in about 150 words according to the following key words.

Key Words: *fault diagnosis*, *direct-vision*, *alarm function*, *replacing parts*, *checking parameters*, *principle analysis*, *PLC technology*, *SKF Equipment Condition Monitor*, *features*

Work Sheet

Part C Broaden Your Horizon

Check for the failures according to the following procedures(Fig. 3-2-5):

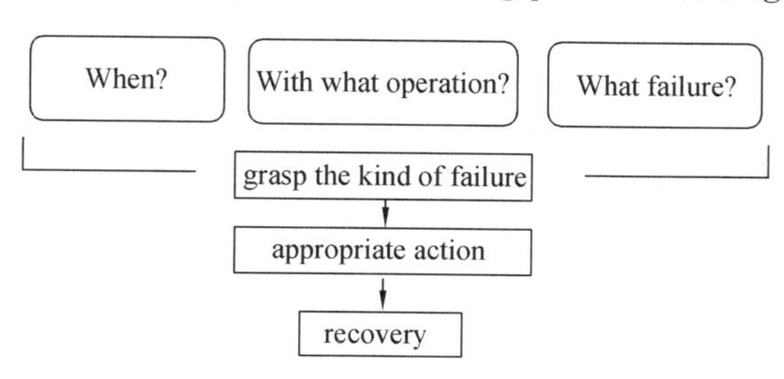

Fig. 3-2-5 Corrective Action for Failures

(1) When did the failure occur?

- Date and time?
- Occurred during operation?
- Occurred when the power was turned on?
- Were there any lightening surges, power failure or other disturbances to the power supply?

(2) How many times has it occurred?

- Only once?
- Occurred how many times? (How many times per hour, per day or per month?)

(3) What operation did it occur with? Which running mode? (automatic operation, manually operation, operating procedure, preceding and succeeding operations, during I/O operation, machine system status, during tool change)

(4) What failure occurred?

任务2 数控机床故障诊断技术

1. 故障诊断方法

当数控机床发生故障时，查找真正的原因并采取恰当的措施非常重要。为此，通常采用下列方法：

(1)直观法

- 问

询问操作人员故障发生时的操作状态以及故障发生的整个经过，弄清楚是突发的还是渐发的，以及该机床以前是否发生过类似的情况等。

◎看

通过观察主驱动轴的转速、冷却液的颜色、被加工工件的损伤情况、传动轴和滚珠丝杠来找到故障发生的位置。

◎听

通过听机床运行时异常响声发生的部位，判断是否是齿轮联轴器、滚珠丝杠螺母副等处发生的故障。

◎摸

触摸机床来判断机床是否过热或者有振动。

◎闻

烧焦味能够帮助判断机床是否受到了剧烈摩擦的影响或电气元件是否被烧毁。

(2)报警功能法

每台数控机床都有报警功能。数控机床的报警功能可以帮助人们判断机床是否存在故障。表 3-2-1 列出了部分数控机床的报警信息。

表 3-2-1　　数控机床的报警信息

序号	信息	故障内容	处理
010	不正确的 G 代码	使用了不能使用的 G 代码或指定了无此功能的 G 代码	修改程序
073	程序号已经被使用	被指定的程序号已经被使用	改变程序号或删除不用的程序，重新执行程序存储
100	参数写入无效	在参数(设置)屏幕上，PWE(参数写入有效)被设置为 1	将 PWE 设置为 0，然后重启系统
150	非法刀具组号	刀具组号超出最大允许值	修改程序
180	通信错误(远程缓冲)	发生远程缓冲连接报警	确认电缆号、参数及 I/O 设备

(3)功能程序测试法

通过数控系统的 GMST 功能编制一个功能测试程序来确定运动的正确性和精度。

(4)备件置换法

如果怀疑某个零件有故障，可以在更换该零件前用备件来替换，以确定机床是否发生故障。

(5)参数检测法

系统参数丢失和不正确的设置都会引起机床性能改变或其他故障。因此，一旦机床发生故障，要及时检查并修正系统参数。

(6)原理分析法

用工作原理来分析故障并找到数控机床发生故障的原因。

2. 基于 PLC 技术的数控机床故障诊断与处理

PLC 是现代数控系统中的一个重要部件，它是联系数控系统和机床的桥梁。为了及时排除与 PLC 相关的故障，需要综合使用多种方法来分析故障，逐步缩小故障范围，并最终排除故障。

(1)根据控制对象的工作原理诊断故障

数控机床的 PLC 程序是按照控制对象的工作原理来设计的。通过分析控制对象的工作原理及 PLC 的 I/O 状态来进行故障诊断是一种非常有效的方法。举例如下：

◎**例 1**:对于配备了 FANUC OTC 系统的数控车床来说,当踩下脚踏开关时,工件不能被装夹。

• **故障分析**:根据机床工作原理,当第一次踩下脚踏开关时,工件应被装夹;当再次踩下脚踏开关时,工件应被松开。使脚踏开关与 PMC 显示屏连接,当踩下脚踏开关后,发现输入 $X2.2$ 始终为零,因此我们可以判断脚踏开关有故障。

• **故障处理**:检查脚踏开关,更换一个新的。

(2)根据 PLC 的 I/O 状态诊断故障

在数控机床中,输入/输出信号的传递一般都要通过 PLC 的 I/O 接口来实现,因此,许多故障都会反映在 PLC 的 I/O 接口上。按照这种特点来诊断故障是十分方便的。如果不是数控系统硬件故障,就可以不必查看梯形图和电路图,直接检查 PLC 的 I/O 接口来找出故障原因。举例如下:

◎**例 2**:某数控机床的防护门不能关闭,机床不能进行自动加工。

• **故障分析**:防护门的打开通过气缸来实现,其关闭由电磁阀 YV2.0 控制,而该电磁阀由 PLC 输出接口 Q2.0 控制。检查发现 Q2.0 的状态为"1",但是电磁阀 YV2.0 不得电,于是判断出中间继电器 KA2.0 损坏,这是因为 PLC 输出接口 Q2.0 是通过它来控制电磁阀 YV2.0 的。

• **故障处理**:更换一个新的中间继电器,更换后故障被排除。

(3)根据报警信息诊断故障

现代数控系统具有丰富的自诊断功能,能在 CRT 上显示故障报警信息。只要我们理解了报警信息的内容,数控设备的故障就能迅速而准确地被排除。举例如下:

◎**例 3**:一台配备了 SINUMERIK 820 数控系统的数控加工中心产生 7035 号报警,报告中的相关信息为工作台分度盘不回落。

• **故障分析**:在 SINUMERIK 810/820 数控系统中,"7"开头的报警为 PLC 操作信息或机床厂设定的报警。处理方法是按照故障信息,调出 PLC 输入/输出状态与副本清单进行对照。工作台分度盘是否回落是由工作台下面的开关 SQ25、SQ28 来检测的。SQ28 检测工作台分度盘是否旋转到位,对应 PLC 输入接口 I10.6。SQ25 检测工作台分度盘是否回落到位,对应 PLC 输入接口 I10.0。工作台分度盘的回落是通过输出接口 Q4.7,由继电器 KA32 驱动电磁阀 YV06 来完成的。从 PLC STATUS 中观察:如果 I10.6 为"1",则表明工作台分度盘旋转到位;如果 I10.0 为"0",则表明工作台分度盘未回落。再观察 Q4.7,如果其为"0",则继电器 KA32 不得电,电磁阀 YV06 不动作,故因工作台分度盘不回落而产生报警。

• **故障处理**:手动移除电磁阀 YV06,观察工作台分度盘是否回落,以判断故障是发生在输出回路还是发生在 PLC 内部。

3. 故障诊断设备——SKF 设备状态监测仪

(1)概述

现在不论是初学者还是专家,通过使用 SKF 设备状态监测仪,都可以轻松、快速并准确地检查旋转设备的状态。

SKF 设备状态监测仪可以进行设备振动速度总值读数测量,而且它能将该测量值与预置于仪器里的 ISO 振动标准进行自动比较。当测量值超出了设置标准,就会有一个"警告"或"危险"的警示显示。同时,它可以进行"加速度包络"测量,所测的包络值会与内置的轴承振动标准进行比较，以判定是否符合标准或指出潜在的轴承故障。SKF 设备状态监测仪还能利用红外传感器进行温度测量，以查出不正常的发热部位。

(2)特征

- 设置快速，易于使用。
- 质量轻、小巧便携且采用人性化设计。
- 报警和危险提示提供了较高的诊断可信度。
- 同时测量振动速度、加速度包络和温度，节省了时间。
- 高效、经济且环保，可充电的 SKF 设备状态监测仪每次充电可使用 10 小时。
- 可选的外部传感器被用于难接触的位置。

学习启示

坚守在平凡的岗位上，勤学苦练，永不懈怠，踏踏实实做好每一件事，你将会把“不可能”变成“可能”。

Module 4
Foundation of CNC Machining Technology

Task 1 Foundation of Mechanical Engineering

1. Metals

(1) Type of Metals

Metals are divided into two general groups: ferrous metals and nonferrous metals. The major types of ferrous metals are iron, carbon steels, alloy steels and tool steels.

The three primary types of cast iron are gray cast iron, white cast iron, and malleable iron. Gray cast iron is primarily used for cast frames, automobile engine blocks, handwheel and cast housings. White cast iron is used for parts such as train wheels. Malleable cast iron is used for tools such as pipes and wrenches.

The three principal types of carbon steel used in industry are low, medium, and high carbon steel. Typical uses of low carbon steel include chains, bolts, nuts, and pipes. Medium carbon steel is used for parts such as gears, crankshafts, machine parts and axles. High carbon steel is used for parts such as files, knives, drills, and razors.

Alloy steels are basically carbon steels with elements added to modify or change the mechanical properties of the steel.

Tool steels are a special grade of alloy steels used for making a wide variety of tools.

The major families of nonferrous metals, indude aluminum and aluminum alloys, copper and copper alloys, magnesium and magnesium alloys, titanium and titanium alloys.

(2) Properties of Metals

The properties of metals are the characteristics that determine how the metal will react under varying conditions. The two principal types of properties are physical and mechanical properties. Physical properties are those fixed properties that are determined naturally and cannot be changed, such as weight, mass, color, and density. Mechanical properties are those properties of metals that can be changed or modified to meet a particular requirement, such as strength, hardness, wear resistance, toughness, plasticity, and ductility.

(3) Heat Treatment of Metals

Heat treatment is a process of controlled heating and cooling of a metal to achieve characteristics change in the properties. The four common heat-treating operations performed on steels are annealing, normalizing, hardening, and tempering.

Purposes of annealing: remove hardness, increase malleability, increase ductility, improve machinability, refine grain structure.

Purposes of normalizing: relieve stresses, produce normal grain size and structure, place steels in the best condition for machining, lessen distortion in heat treating.

Purposes of hardening: increase hardness, strength and wear resistance.

Purposes of tempering: reduce hardness to desired level, increase shock resistance and impact strength, reduce brittleness, relieve stresses caused by rapid cooling.

2. Engineering Drawing

Typical drawings in machine manufacturing are classified as part drawings and assembly drawings.

(1) Part Drawing

Part drawings are frequently used as instructing for manufacture and inspecting for the parts. An integrated part drawing should include a set of drawings, overall dimensions, necessary technical requirements and full contents of title block.

(2) Assembly Drawing

Assembly drawings are used in explaining machines or components. In mechanical design, the designer first draws the assembly drawing to specifically express the working principle and structure of the designed machine or components, and then draws the part drawing according to the assembly drawing. In the process of mechanical manufacturing, the parts are first processed according to the part drawing, and then assembled into components or machines according to the assembly drawing.

3. Machine Elements

(1) Gears

The most common example of a machine element is a gear. There are several kinds of gears used in modern machinery. Some of those are spur gears, helical gears, racks, bevel gears, worm gears and so on.

Spur gears (Fig. 4-1-1) are the most widely used style of gears and are used to transmit rotary motion between parallel shafts, while maintaining uniform speed and torque.

Helical gears (Fig. 4-1-2) are similar to spur gears with the exception that the teeth are cut at an angle to the axis of the shaft—the helix angle. The helix cut creates a wider contact area enabling higher loads and torques to be achieved.

Bevel gears (Fig. 4-1-3) are used solely to transmit rotary motion between intersecting shafts.

Fig. 4-1-1 Spur Gears

Fig. 4-1-2 Helical Gears

Fig. 4-1-3 Bevel gears

(2) Belt Drives (Fig. 4-1-4)

The belt drive has the advantages of stable transmission, simple structure and low cost, which is suitable for the occasions with long center distance between two shafts.

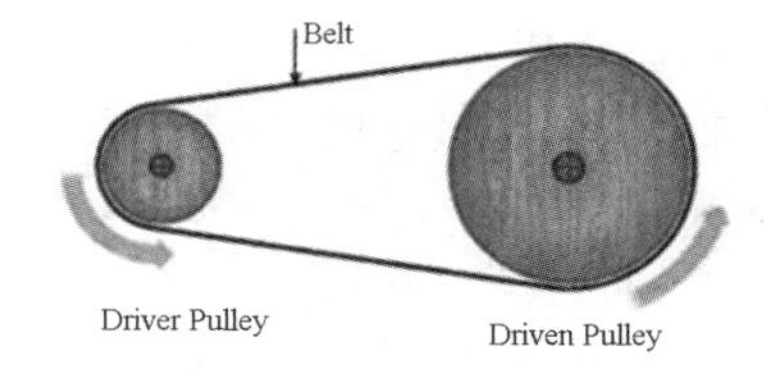

Fig. 4-1-4 Belt Drive

There are four main types of belt drives: flat belt, V belt, circular belt and synchronous belt. The flat belt is capable of transmitting power over long center distances between pulley centers and also widely used in high-speed belt drive. The V belt is much higher than that of flat belt, and it is widely used in power transmission. The circular belt is generally used for low speed and light load transmission. The synchronous belt has evenly spaced teeth on the inside circumference. It does not slip and hence transmits power at a constant velocity ratio.

(3) Chain Drives

Chain drives provide almost any speed ratio. Their chief advantage over gear drives is that they can be used with arbitrary center distances. The most familiar chain drive is the roller chain drive on a bicycle.

Translation

模块 4 数控加工技术基础

任务 1 机械工程基础

1. 金属材料

(1)金属材料的类型

金属材料分为两类:铁类金属和非铁金属。铁类金属主要有铸铁、碳素钢、合金钢和工具钢。

铸铁主要分为三种:灰铸铁、白口铸铁和可锻铸铁。灰铸铁主要用于制造支架、汽车发动机缸体、手轮和壳体。白口铸铁用来制造如火车车轮之类的零件。可锻铸铁可用来制造如管道和扳手之类的零件。

工业上使用的碳素钢主要有三种:低碳钢、中碳钢和高碳钢。低碳钢主要用于制造链条、螺钉、螺母和管道。中碳钢用于制造齿轮、曲轴、机器零件和车轴。高碳钢用于制造锉刀、刀、钻头、剃刀。

合金钢是在碳素钢中加入一定的合金元素以改变或提高其力学性能的钢铁材料。

工具钢是一种特殊等级的合金钢,用来制造各种工具的钢。

常用非铁金属有铝及铝合金、铜及铜合金、镁及镁合金、钛及钛合金。

(2)金属材料的性能

金属材料的性能是指金属在不同条件下的反应特性。金属材料的性能主要分为两类:物理性能和力学性能。物理性能是决定于材料本身且不能改变的固有性能,如重量、质量、颜色、密度。材料的力学性能可以改变,以满足特定的要求,如强度、硬度、耐磨性、韧性、弹性和延展性。

(3)金属材料的热处理

热处理是通过对金属进行加热和冷却控制,以使其性能得到显著改变的一种工艺方法。钢铁材料最常用的四种热处理方法是:退火、正火、淬火和回火。

退火的目的:降低硬度,提高可锻性,提高延展性,改善机械加工性能,改善组织。

正火的目的:消除应力,细化晶粒和改善结构均匀性,为机械加工做准备,减少热处理产生的变形。

淬火的目的:提高硬度、强度和耐磨性。

回火的目的:降低硬度到所需要的水平,提高抗冲击性和冲击强度,降低脆性,消除快速冷却产生的内应力。

2. 工程图纸

在机械制造中比较典型的机械图样有两种:零件图和装配图。

(1)零件图

零件图是直接指导制造和检验零件的图样,一张完整的零件图包括一组图形、完整的尺寸、必要的技术要求和内容完整的标题栏。

(2)装配图

装配图是表达机器或组成部件的图样,在机械设计中,设计者首先画出装配图,具体表达所设计机器或部件的工作原理和结构,然后根据装配图分别绘制零件图。在机械制造过程中,首先根据零件图加工零件,然后按照装配图装配成部件或组装成机器。

3. 机械零件

(1)齿轮

机械零件中最常用的就是齿轮。在现代机械中使用的齿轮有很多种,其中有直齿圆柱齿轮、斜齿轮、齿条、锥齿轮、蜗轮蜗杆等。

直齿圆柱齿轮(图 4-1-1)是齿轮中应用最广泛的一种类型,它用于在平行轴间传递旋转运动,并且保持恒定的速度和转矩。

斜齿轮(图 4-1-2)与直齿圆柱齿轮相似,只是斜齿轮的轮齿与轴的中心线呈一定角度-螺旋角,这样可以产生更大的接触面积,从而可以承受更高的载荷和扭矩。

锥齿轮(图 4-1-3)仅用于两相交轴间传递旋转运动。

图 4-1-1 直齿圆柱齿轮

图 4-1-2 斜齿轮

图 4-1-3 锥齿轮

(2)带传动(图 4-1-4)

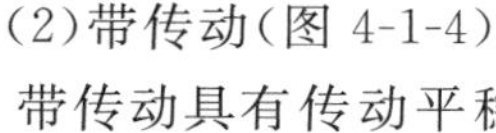

带传动具有传动平稳,结构简单、成本低廉等优点,适用于两轴中心距较大的场合。

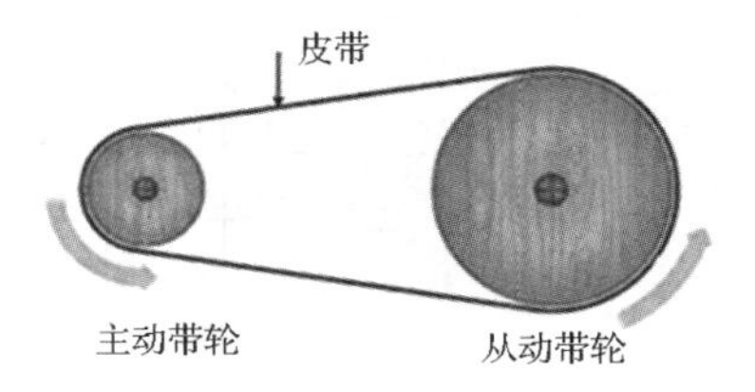

图 4-1-4 带传动

带传动主要有四种类型:平带、V 带、环形带和同步带。平带常用于带轮中心距较大的场合,也广泛用于高速带传动。V 带传动的功率远高于平带传动,广泛用于动力传动。圆形带一般用于低速轻载传输。同步带的内圆周上有均匀分布的齿,它不会打滑,因此能够以恒定的速比传动。

(3)链传动

链传动几乎可以提供任何速比。与齿轮传动相比,它的主要优点是链传动可用于任意中心距。最常见的是自行车上的滚子链传动。

Task 2 Foundation of Control Technology

1. Electrical Control

(1) The Working Principles of Electrical Control

Electrical control refers to the use of electrical logic relations and operations to complete the control of the object of the task, that is, to achieve a certain purpose, to apply the required operations.

(2) The Composition of Electrical Control System

The electrical control system includes electrical control components (such as contactor), electrical protection components (such as fuse, thermal relay, etc.), electric actuator (such as 3-phase asynchronous motor), electric circuit (such as main circuit, control circuit etc.), mechanical transmission device (such as belt and reducer, etc.).

(3) The Applied Examples of Electrical Control

The electrical schematic is expressed by graphics and text symbols which represent the connection between electrical components and the electrical working principle in the circuits, but it does not reflect the actual size of the electrical components and installation location. The electrical schematic of KH-CA6140 lathe is shown as Fig. 4-2-1. It can be seen from the figure that M1, M2 and M3 are three three-phase asynchronous motors, and contactors KM1, KM2 and KM3 control the start and stop of the three motors respectively. SB1 is the stop button. SB2 is the start button. After the power is turned on, press SB2, the coil of contactor KM1 is powered on, the auxiliary normally open contact and main contact of KM1 are closed, and the spindle motor M1 starts to run.

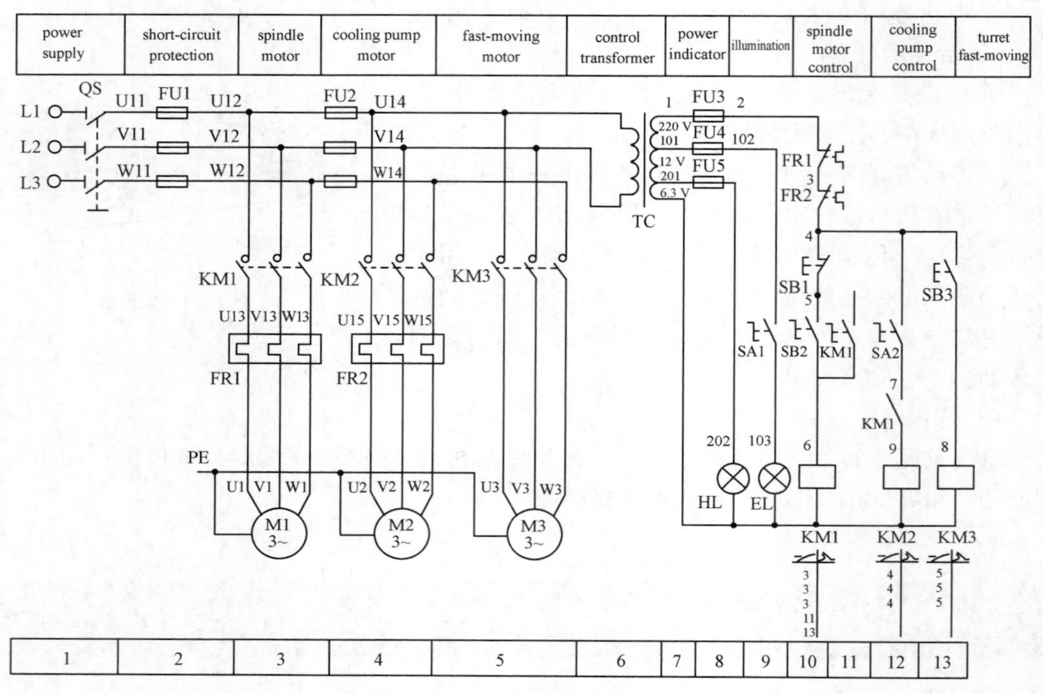

Fig. 4-2-1 Electrical Schematic of KH-CA6140 Lathe

2. Hydraulic Control

(1) The Working Principles of Hydraulics Transmission

The hydraulic transmission is a transmission form of transfer and control energy to use the liquid as the transmission medium, which can realize the transmission and automatic control of all kinds of machinery.

(2) The Composition of Hydraulic Transmission System

The hydraulic transmission system includes energy devices (such as hydraulic pump), actuating elements (such as hydraulic cylinder, hydraulic motor), control elements (such as relief valve, directional control valve, flow control valve, etc.), auxiliary elements (such as reservoir, filter, energy accumulator, etc.), working medium (such as hydraulic oil).

(3) The Applied Examples of Hydraulic Transmission

In order to simplify the representation of hydraulic transmission systems, graphical symbols are usually used to draw the schematic diagram of the system. Fig. 4-2-2 shows the hydraulic system of the grinding machine workbench. While working, the relief valve is used to adjust or limit the pressure of the hydraulic system, flow control valve can adjust the size of the hydraulic cylinder speed, and three-position four-way directional control valve can make the workbench move or stop.

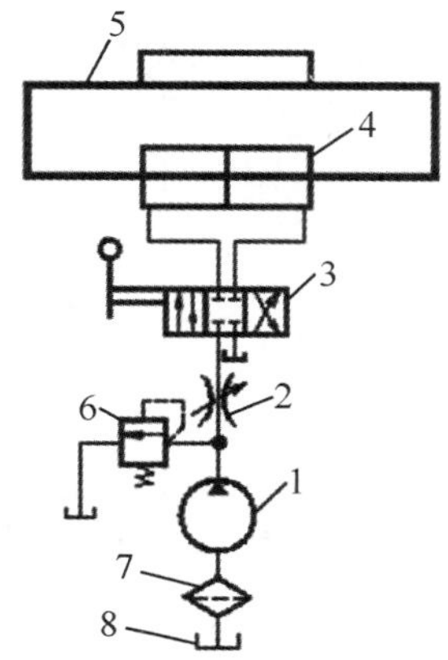

Fig. 4-2-2 Hydraulic System of the Grinding Machine Workbench

1—Hydraulic Pump; 2—Flow Control Valve; 3—Directional Control Valve; 4—Hydraulic Cylinder; 5—Workbench; 6—Relief Valve; 7—Filter; 8—Reservoir

3. Pneumatic Control

(1) The Working Principles of Pneumatic Transmission

The pneumatic transmission is a transmission form which uses compressed air as transmission medium to transfer and control energy or signal, and is also a technology to realize the automation of production process by controlling and driving of all kinds of machinery and equipment.

(2) The Composition of Pneumatic Transmission System

The pneumatic transmission system includes air supply devices, pneumatic actuator elements (such as air cylinder, air motor), pneumatic control elements (such as pneumatic pressure reducing valve, throttle valve, reversing valve), pneumatic auxiliary elements (such as pipe, pressure gauge, filter, muffler, oil mist device, etc.), working medium (such as compressed air).

(3) The Applied Examples of Pneumatic Transmission

The principle of pneumatic control for a tool changer is shown in Fig. 4-2-3. In the process of tool change, the system can achieve positioning of the spindle, tool releasing, tool drawing, blowing into the spindle hole, etc..

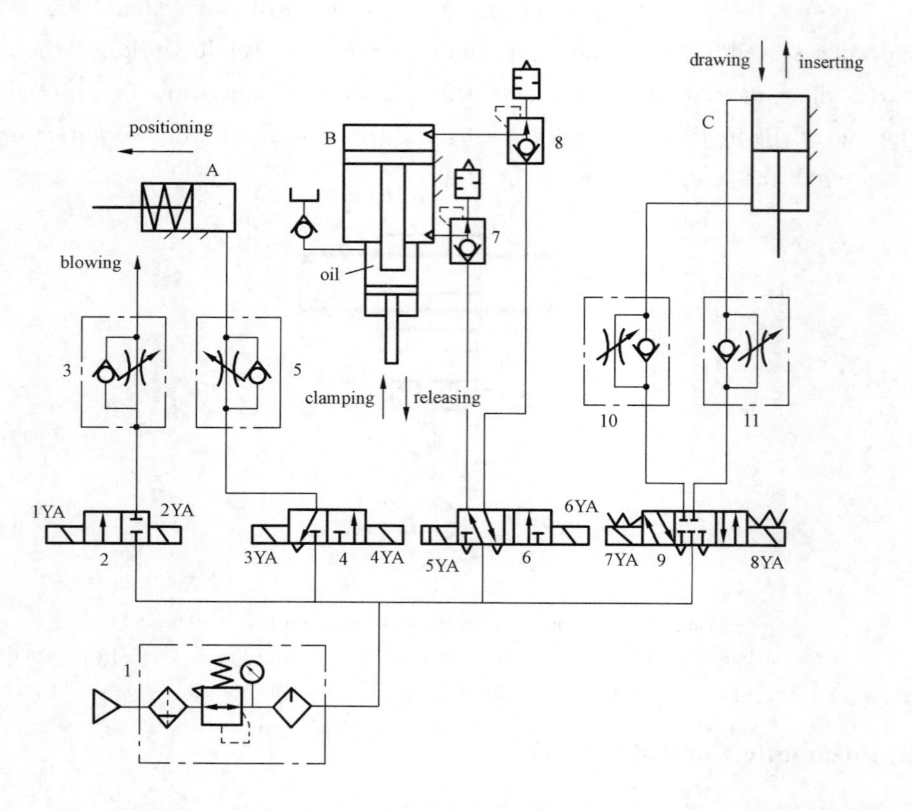

Fig. 4-2-3 Principle of Pneumatic Control for a Tool Changer

1—pneumatic FRL; 2—2/2-solenoid valve; 3, 5, 10, 11—throttle valve; 4—3/2-solenoid valve; 6—5/2-solenoid valve; 7, 8—fast bleeder valve; 9—5/3-solenoid valve

4. PLC

(1) The Working Principles of PLC

PLC is the abbreviation of programmable logic control. PLC is a kind of general industrial control computer. PLC's work is a continuous cycle scanning process. The whole scan process includes five stages: internal processing, communication processing, input scan, execute user program and output processing.

(2) The Composition of PLC

PLC is composed of hardware and software. The hardware consists of five parts: central processing unit(CPU), memory, input/output unit, power unit, programming unit.

(3) The Applied Examples of PLC

PLC ladder diagram of CNC machine lubrication system is shown in Fig. 4-2-4. The system can realize the monitoring of the starting operation of the lubricating motor and the failure of the lubricating system.

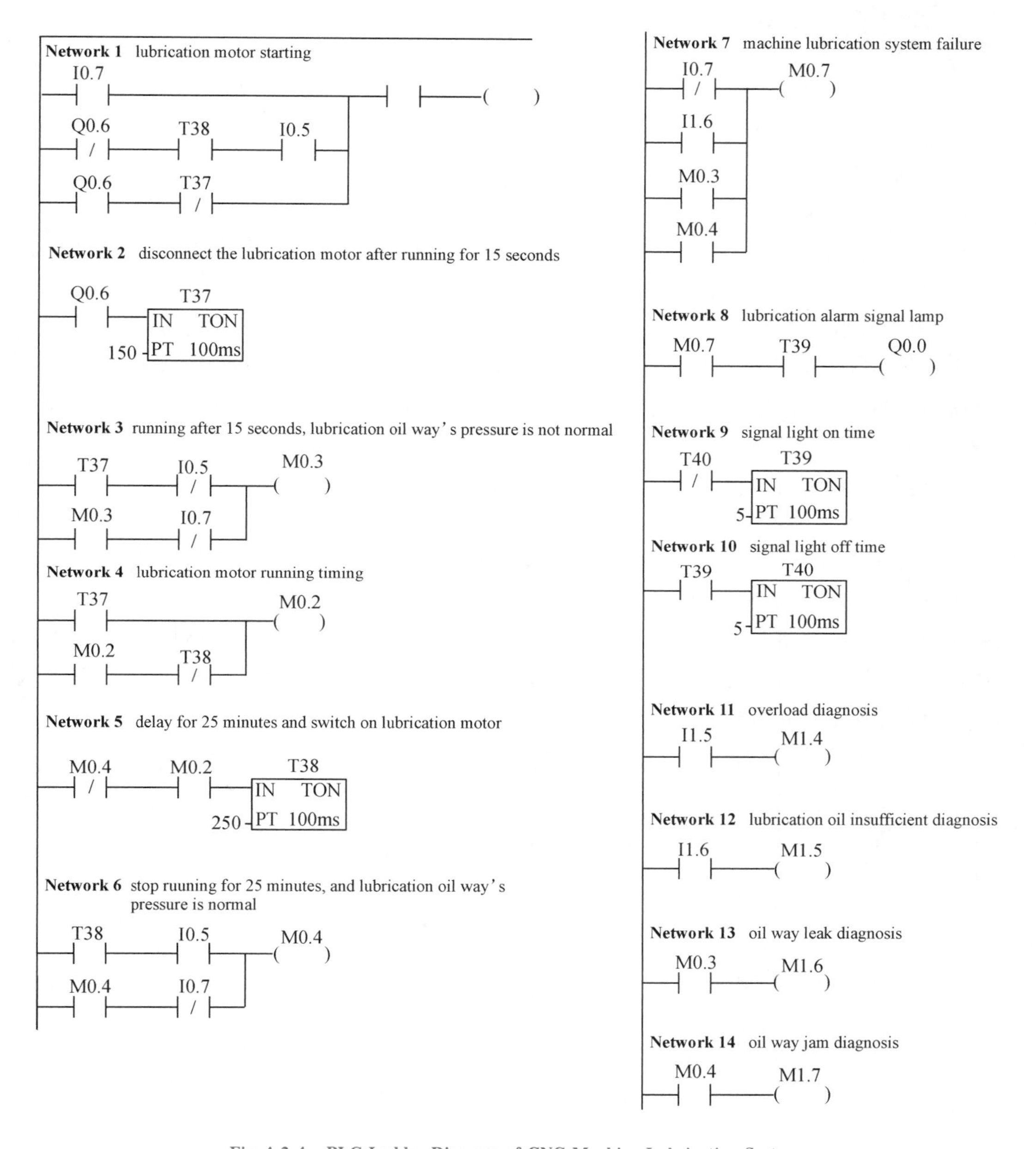

Fig. 4-2-4 PLC Ladder Diagram of CNC Machine Lubrication System

Translation

任务2 控制技术基础

1. 电气控制

(1)电气控制的工作原理

电气控制是指利用电气逻辑关系和运算完成对被控对象的控制任务，即为达到某种目的，对被控对象施加所需的操作。

(2)电气控制系统组成

电气控制系统包含电气控制元件(如接触器)、电气保护元件(如熔断器、热继电器等)、电气执行元件(如三相异步电动机)、电气线路(如主电路、控制电路等)、机械传动装置(如皮带、减速器等)。

(3)电气控制应用实例

电气原理图是用图形和文字符号来表达的，这些图形和文字符号表示电路中各个电气元件的连接关系和电气工作原理，但它并不反映电器元件的实际大小和安装位置。图4-2-1所示为KH-CA6140型车床的电气原理图。从图4-2-1中可以看出，M1、M2、M3为三台三相异步电动机，接触器KM1、KM2和KM3分别对三台电动机进行启停控制。SB1为停止按钮，SB2为启动按钮。接通电源后，按下SB2，接触器KM1线圈得电，KM1辅助常开触点和主触点闭合，主轴电动机M1启动运行。

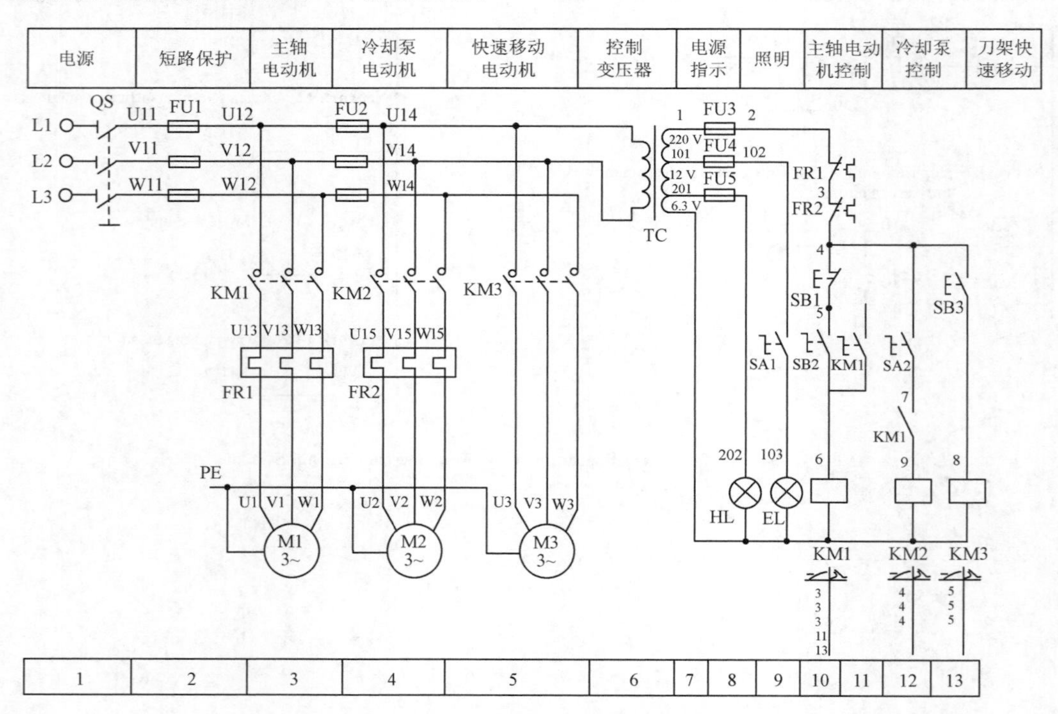

图4-2-1 KH-CA6140型车床的电气原理图

2. 液压控制

(1)液压传动的工作原理

液压传动是利用液体作为传动介质对能量进行传递和控制的一种传动形式,从而实现各种机械的传动和自动控制。

(2)液压传动系统的组成

液压传动系统包括能源装置(如液压泵)、执行元件(如液压缸、液压马达)、控制元件(如溢流阀、方向控制阀、流量控制阀等)、辅助元件(如油箱、过滤器、蓄能器等)、工作介质(如液压油)。

(3)液压传动应用实例

为了简化液压传动系统的表示方法,通常采用图形符号来绘制系统原理图。图 4-2-2 所示为磨床工作台的液压系统。工作时,溢流阀用来调节或限制系统的压力,流量控制阀可以调节液压缸运动速度的大小,三位四通方向控制阀可实现工作台的运动和停止。

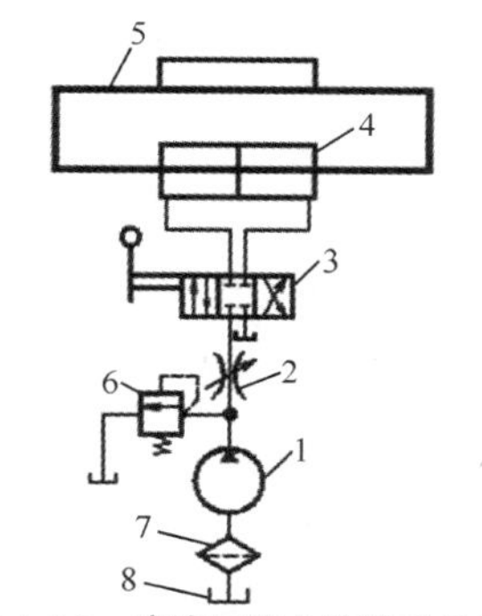

图 4-2-2 磨床工作台的液压系统

1—液压泵;2—流量控制阀;3—方向控制阀;4—液压缸;5—工作台;6—溢流阀;7—过滤器;8—油箱

3. 气动控制

(1)气压传动的工作原理

气压传动以压缩空气为工作介质,进行能量或信号传递及控制的一种传动方式,也是控制和驱动各种机械和设备,以实现生产过程自动化的技术。

(2)气压传动系统的组成

气压传动系统包括气源装置、气动执行元件(如气缸、气马达)、气动控制元件(如气动减压阀、节流阀、换向阀等)、气动辅助元件(如管道、压力表、过滤器、消声器、油雾器等)、工作介质(如压缩空气)。

(3)气动控制应用实例

换刀装置的气动控制原理如图 4-2-3 所示。该系统在换刀过程中能够实现主轴定位、松刀、拔刀、向主轴孔吹气等动作。

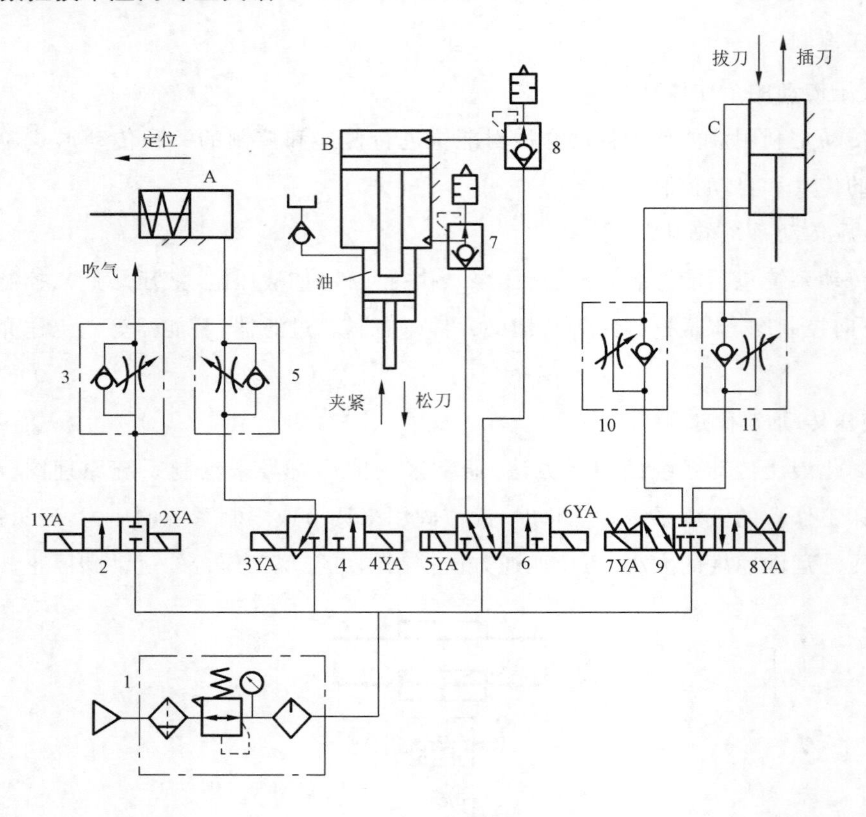

图 4-2-3 换刀装置的气动控制原理

1—气动三联件；2—两位两通电磁换向阀；3、5、10、11—单向节流阀；4—两位三通电磁换向阀；6—两位五通电磁换向阀；7、8—快速排气阀；9—三位五通电磁换向阀

4. PLC

(1)PLC 的工作原理

PLC 是可编程控制器的简称,PLC 是一种通用工业控制计算机。PLC 的工作是周而复始地循环扫描的过程。整个扫描工作过程包括内部处理、通信处理、输入扫描、用户程序执行、输出处理五个阶段。

(2)PLC 的组成

PLC 由硬件和软件两部分组成。PLC 硬件部分通常由中央处理单元、存储器、输入/输出单元、电源单元、编程器五部分组成。

(3)PLC 应用实例

图 4-2-4 为数控机床润滑系统的 PLC 梯形图。该系统可以实现润滑电动机启动运行和润滑系统出现故障时的监控。

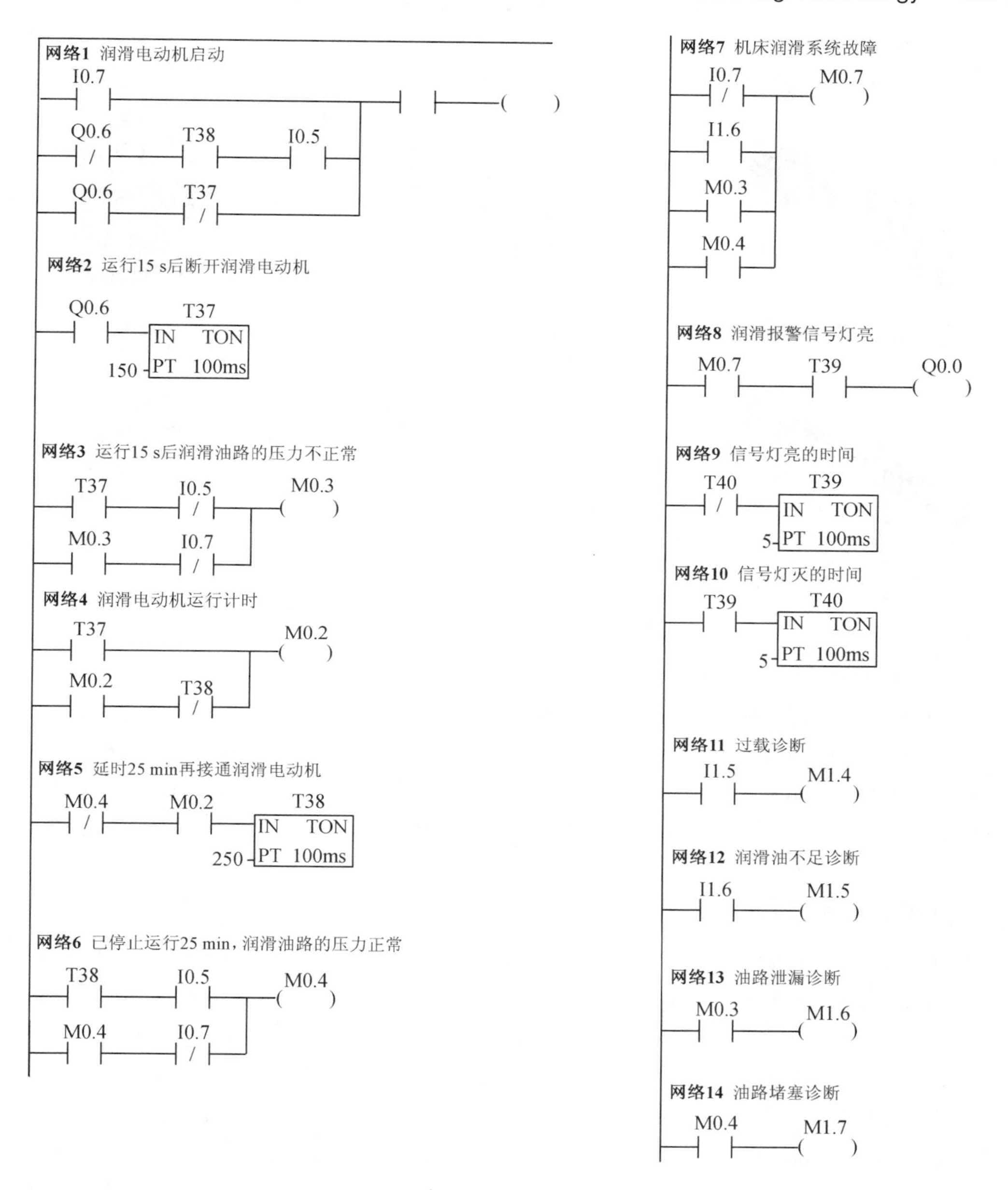

图 4-2-4 数控机床润滑系统的 PLC 梯形图

Task 3 Foundation of CNC Turning Process

1. Cutting Tools for Turning

(1) Types of Turning Tools

The commonly used turning tools are as follows: external turning tool

(Fig. 4-3-1(a)), groove tool(Fig. 4-3-1(b)), thread tool(Fig. 4-3-1(c)), internal turning tool(Fig. 4-3-1(d)), twist drill(Fig. 4-3-1(e)) and center drill(Fig. 4-3-1(f)), etc..

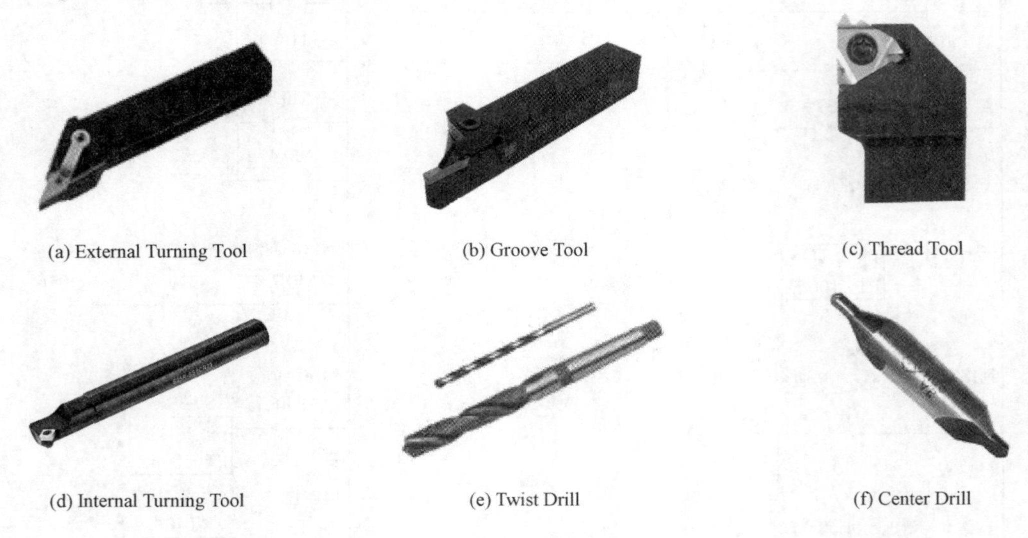

(a) External Turning Tool (b) Groove Tool (c) Thread Tool

(d) Internal Turning Tool (e) Twist Drill (f) Center Drill

Fig. 4-3-1 Types of Turning Tools

(2) Applications of Turning Tools

External turning tool is used for turning cylinder, taper and facing surface. Groove tool is used in parting or cutting off the workpiece. Thread tool is used to cut a standard 60-degree thread. Internal turning tool is used in a boring operation. Twist drill is the most common type of drill in rough drilling operation. Center drill is used to provide positioning for drilling operation.

(3) Material and Property of Cutting Tools

Cutting tools must possess certain mechanical properties in order to function adequately during the cutting operations. These properties include high hardness, toughness, abrasion resistance and the ability to withstand high pressure.

The commonly used turning tools are made of high-speed steel and cemented carbide.

High-speed steel(HSS) is a kind of alloy steel that contains a certain percentage of alloying elements, such as tungsten(18%), chromium(4%), molybdenum, vanadium and cobalt. Tools made of HSS can retain their hardness at an elevated temperature up to 600 ℃.

Cemented carbide is usually composed of tungsten carbide, barium carbide, titanium carbide or tantalum carbide and cobalt in certain combinations. A typical composition of cemented carbide is 85 to 95 percent of carbides of tungsten and the remainder is cobalt. Cemented carbides are the most widely used tool materials in the machining industry.

Coated carbides are coated with wear resistant materials on the surface of cemented carbide insert. Its wear resistance can be improved by 200% to 500%.

(4) Geometry Angles of Cutting Tools

The standard geometry angles of cutting tools are shown in Fig. 4-3-2. For turning tools, the most important angles are the rake angle and the relief angle.

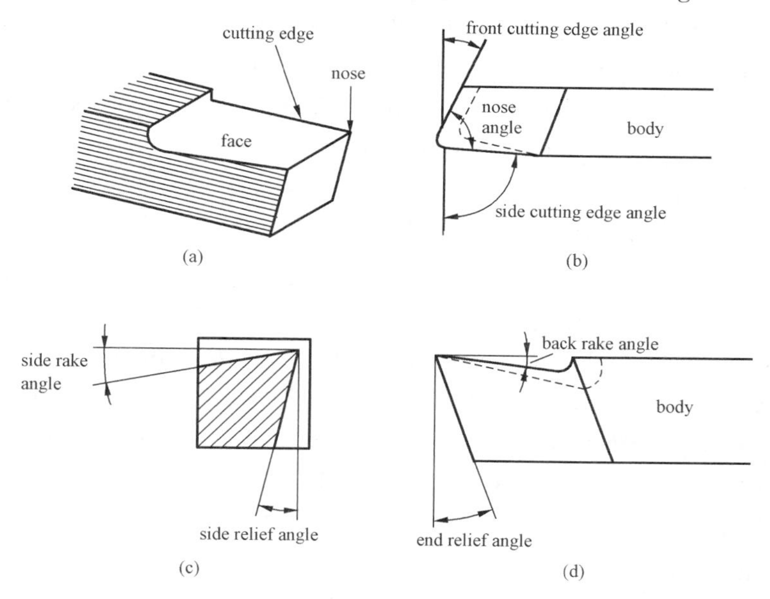

Fig. 4-3-2 Standard Geometry Angles of Cutting Tools

Rake angle can be positive, negative or zero. Positive rake angles reduce the cutting forces resulting in smaller deflections of the workpiece, tool holder and machine. If the rake angle is too large, the strength of the tool will be reduced as well as its capacity to conduct heat. In machining hard materials, the rake angle must be small, even negative for cemented carbide tools. For high-speed steels, the rake angle is normally chosen in the positive range.

Relief angle serves to eliminate friction between the workpiece and the end flank. The degree of relief angle has important effect on surface quality of the workpiece. At the same time, relief angle affects the strength of tool edge.

2. Cutting Dosage

The three primary factors in turning operation are cutting speed, feed rate and depth of cut. Other factors such as kind of material and type of tool have a large influence.

(1) Cutting Speed

For lathe operation, the cutting speed is defined as the rate at which a point on the circumference of the workpiece passes the cutting tool. The relationship between the cutting speed and rpm can be given by the following equation:

$$v=\pi Dn/1\,000$$

Where v—cutting speed(m/min);

D—diameter of the workpiece(mm);

n—rpm(r/min).

Every material has an ideal cutting speed. This is the optimum speed at which the material can be cut safely, in order to obtain a good quality of surface. The cutting speed is dependent primarily upon the material being machined as well as the material of the cutting tools and can be obtained from handbooks provided by cutting tool manufacturers. There are also other variables that affect the optimal value of the cutting speed. These include the tool geometry, the type of lubricant and coolant, the feed rate and the depth of cut, etc..

(2) Feed Rate

The feed rate is the distance the tool advances into the workpiece per revolution. The selection of a suitable feed rate depends upon many factors, such as the required surface quality, the depth of cut and the geometry of the tool used. Finer feed rate produces better surface quality, whereas higher feed rate reduces the machining time. Therefore, it is generally recommended to use higher feed rate for roughing operation and finer feed rate for finishing operation. Recommended values for feed rate, which can be found in handbooks provided by cutting tool manufacturers, will only be taken as guidelines.

(3) Depth of Cut

Depth of cut is defined as the distance that the cutting tools are plunged into the workpiece. It is typically measured in millimeters. For turning operation, the depth of cut can be calculated by the following equation(Fig. 4-3-3):

$$a_p = (d_w - d_m)/2$$

Where a_p—depth of cut;

d_w—the diameter of the new surface;

d_m—the diameter of the finished surface.

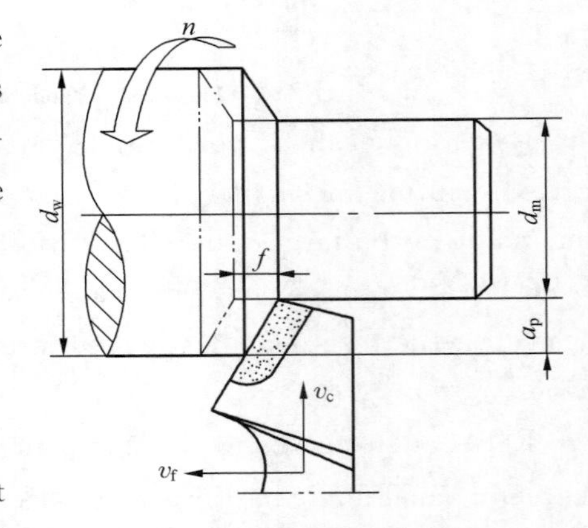

Fig. 4-3-3 Depth of Cut

To determine the depth of cut we must first select the proper cutting tool, the proper machine and a suitable setup. The depth of cut directly influences the tool life. If the depth of cut is too large, the insert may be overloaded, causing immediate breakage. The handbooks provide the recommended ranges for depths of cut for each insert. In the finishing operation, it is important to select a small depth of cut.

3. An Example of Process Analysis of Turning

In Fig. 4-3-4, we can see that the workpiece is a rotational component and a lathe should be used. The workpiece needs to be machined by some of the common turning operations—facing cut, outside surface turning, grooving, threading and drilling, etc.. The surfaces to be machined are shown in Fig. 4-3-5.

S1,S11:facing surface;

S3,S6,S8,S10:cylindrical surface;

S2,S7:arc surface;

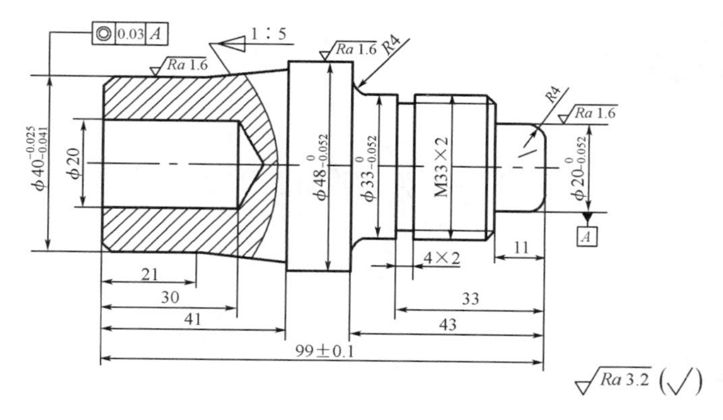

Fig. 4-3-4 Part Drawing

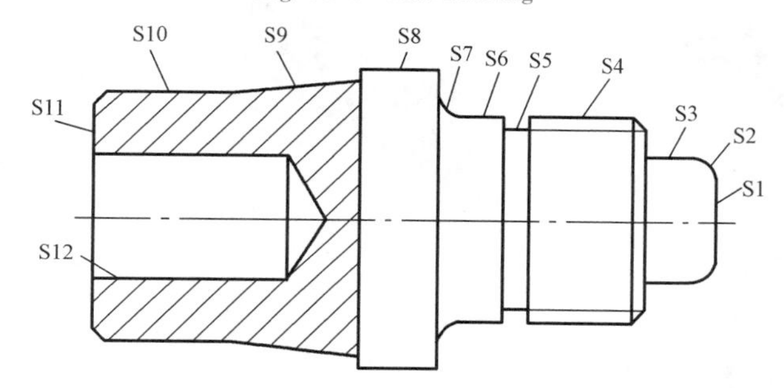

Fig. 4-3-5 Surfaces to Be Machined

S9: taper surface;

S4: cylinder thread surface;

S5: thread relief with 4×2;

S12: 20 diameter hole.

Analyzing the part drawing, we know that S1, S2, S3, S6, S7, S8, S9, S10 and S11 require outside surface turning, S4 requires turning followed by threading, and drilling operations are needed for S12. The following is the sequence of machining the workpiece:

Setup 1:

- Chuck the workpiece with three-jaw chuck;
- Turn S1, S2, S3, S4, S6, S7, S8;
- Cut the 4×2 thread relief;
- Thread S4.

Setup 2:

- Chuck the workpiece on S8;
- Turn S11, S10 and S9;
- Drill center hole and S12 hole.

Translation

任务3 数控车削加工工艺基础

1. 车削刀具

(1)车刀的类型

常用的车刀有外圆车刀(图 4-3-1(a))、切槽刀(图4-3-1(b))、螺纹车刀(图 4-3-1(c))、内孔车刀(图 4-3-1(d))、麻花钻(图 4-3-1(e))和中心钻(图 4-3-1(f))等。

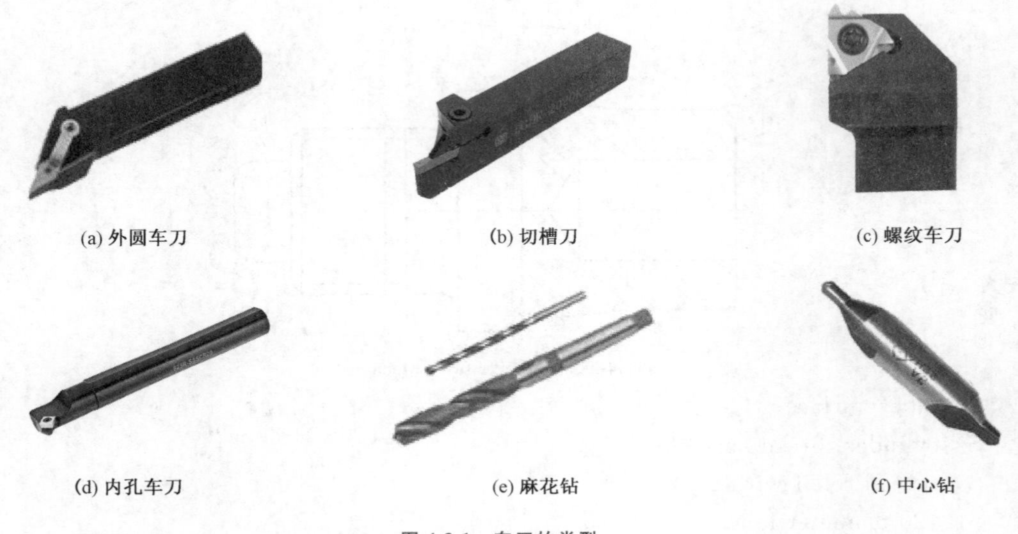

(a) 外圆车刀 (b) 切槽刀 (c) 螺纹车刀

(d) 内孔车刀 (e) 麻花钻 (f) 中心钻

图 4-3-1 车刀的类型

(2)车刀的应用

外圆车刀用于切削圆柱面、锥面和端面;切槽刀用于分割或切断工件;螺纹车刀用于切削 60°标准螺纹;内孔车刀用于加工内孔表面;麻花钻是孔的粗加工操作中最常用的钻头;中心钻为钻孔加工提供定位。

(3)切削刀具的材料及性能

切削刀具必须具有一定的机械性能,以保证在加工过程中能够充分发挥作用。这些性能包括高硬度、韧性、耐磨性以及抗高压的能力。

常用车刀由高速钢和硬质合金制成。

高速钢(HSS)是一种合金钢,它含有一定量的合金元素,如钨(18%)、铬(4%)、钼、钒和钴。用高速钢制造的刀具在温度升高至 600 ℃时仍能保持其硬度。

硬质合金通常由碳化钨、碳化钡、碳化钛或钽以及钴按一定的比例混合而成。典型的硬质合金的组成是 85%~95%的碳化钨,其余是钴。硬质合金是机械加工中使用最广泛的一种刀具材料。

涂层硬质合金是在硬质合金刀片表层镀有一层耐磨材料,其耐磨性可提高 200%~500%。

(4)切削刀具的几何角度

切削刀具的标准几何角度如图 4-3-2 所示。对于车刀来说，最重要的角度是前角和后角。

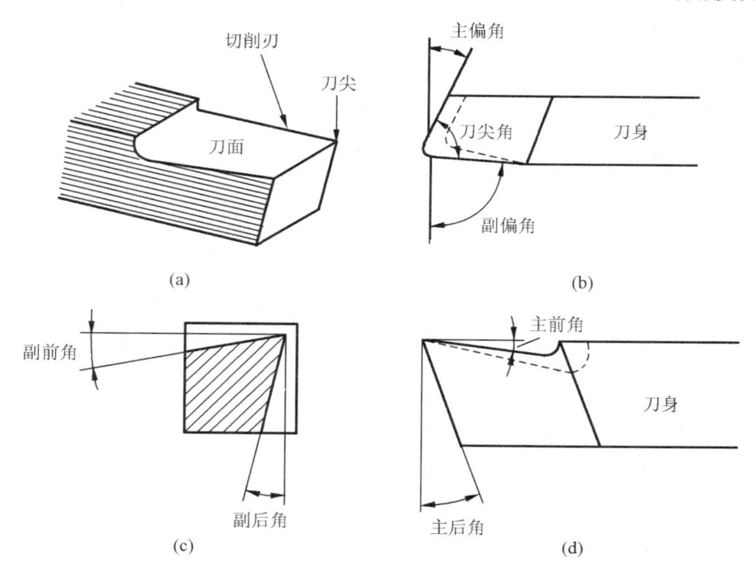

图 4-3-2 切削刀具的标准几何角度

前角可以是正的、负的或者为零。正前角能够减小切削力，使工件、刀柄和机床产生更小的变形。如果前角太大，刀具的强度和散热能力就会降低。在加工硬度大的材料时，前角须很小，对于硬质合金刀具来说，前角甚至是负值。对于高速钢材料，前角通常选择正值。

后角用来降低工件与后刀面之间的摩擦。后角的大小对工件表面质量有很大影响，同时影响刀刃的强度。

2. 切削用量

车削加工的三要素：切削速度、进给量和切削深度。此外，刀具材料和刀具类型对车削加工也有很大的影响。

(1)切削速度

对于车削加工，切削速度定义为刀具经过工件边缘上一点的线速度。切削速度与主轴转速之间的关系如下：

$$v=\pi Dn/1\ 000$$

式中 v ——切削速度(m/min)；

D——工件直径(mm)；

n——主轴转速(r/min)。

每种材料都有一个合理的切削速度，这个速度是保证切削正常进行以获得较好表面质量的最佳速度。切削速度的大小主要取决于被加工材料和切削刀具的材料，这个数值可以从刀具制造商所提供的手册中查到。还有其他一些因素会影响切削速度的最佳值，包括刀具的几何参数、润滑油和冷却液的种类、进给量和切削深度等。

(2)进给量

进给量是指工件每转一周，刀具与工件之间相对移动的距离。选择合适的进给量取决于许多因素，例如加工表面的质量要求、切削深度和所用刀具的几何参数。选取较小的进给量可获得较好的表面质量，选取较大的进给量可减少加工时间。因此，一般情况下常选取较大的进给量进行粗加工，而选取较小的进给量进行精加工。从刀具制造商提供的手册中获得的进给量推荐值仅作为参考。

(3)切削深度

切削深度是指刀具切入工件的深度，一般以毫米为单位。对于车削加工，切削深度的计算公式如下(图 4-3-3)：

$$a_p = (d_w - d_m)/2$$

式中 a_p——切削深度；

d_w——工件待加工表面直径；

d_m——工件已加工表面直径。

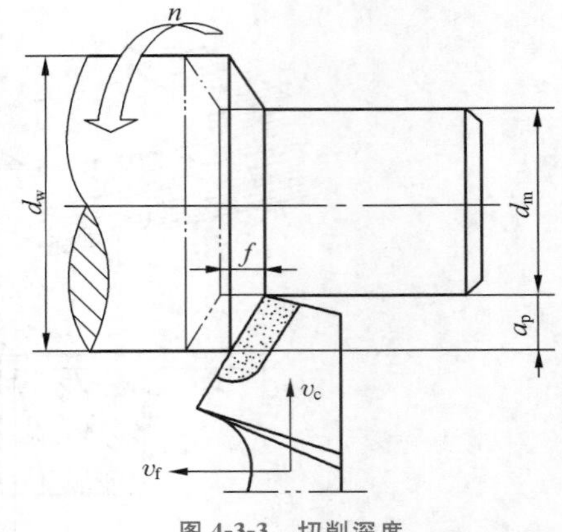

图 4-3-3 切削深度

确定切削深度必须首先选择合适的切削刀具、合适的机床和安装条件。切削深度直接影响刀具的寿命。如果切削深度太大，刀片可能会因过载而被损坏。从手册中可以查到每种刀片的切削深度的推荐范围。精加工时，应选取较小的切削深度。

3. 车削加工工艺分析实例

从图 4-3-4 中可以看出，该零件为回转体，应使用车床加工。该零件需要用一些常用的车削加工操作进行加工——端面车削、外圆车削、切槽、车螺纹以及孔加工等。该零件的被加工表面如图 4-3-5 所示。

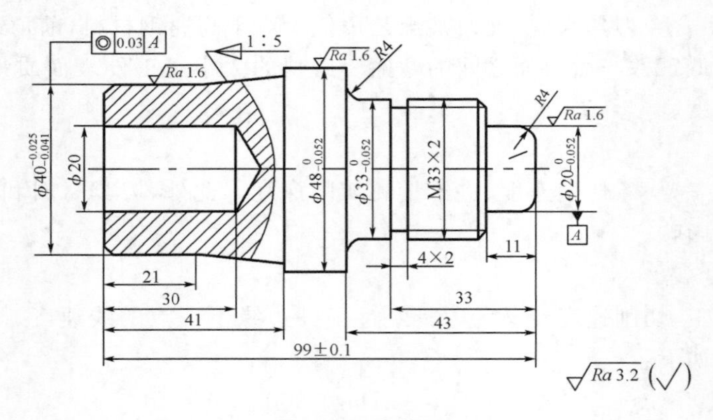

图 4-3-4 零件图

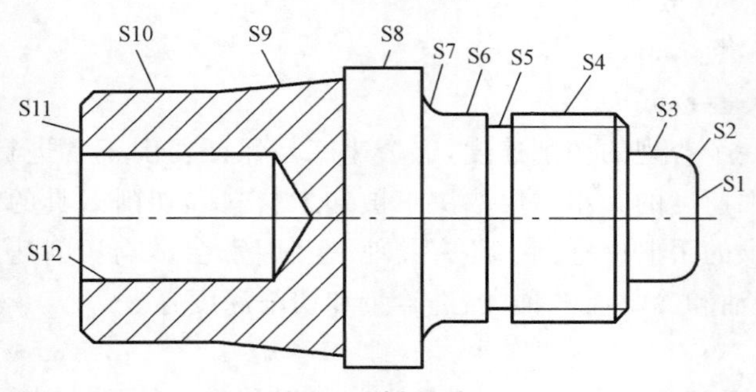

图 4-3-5 被加工表面

S1、S11：端面；

S3、S6、S8、S10：外圆面；

S2、S7：圆弧面；

S9：圆锥面；

S4：外螺纹面；

S5：4×2 螺纹退刀槽；

S12：直径为 20 的孔。

通过分析零件图可知，S1、S2、S3、S6、S7、S8、S9、S10 和 S11 需要外轮廓车削加工，S4 需要车削后进行螺纹加工，S12 需要进行孔加工。零件的加工顺序如下：

第一次装夹：

- 用三爪卡盘装夹工件；
- 车削 S1、S2、S3、S4、S6、S7、S8 表面；
- 切削 4×2 螺纹退刀槽；
- 车螺纹 S4。

第二次装夹：

- 装夹工件 S8 表面；
- 车削 S11、S10 和 S9 表面；
- 钻中心孔后钻孔 S12。

Task 4 Foundation of CNC Milling Process

1. Selection of Tools(Fig. 4-4-1)

The workpiece is clamped on the machine vice with a dial test indicator calibrated. Edge-finder is used for *X*/*Y* tool setting, and *Z*-zero-setter is used for *Z* tool setting.

(a) Machine Vice

(b) Dial Test Indicator

(c) Edge-finder

(d) *Z*-zero-setter

Fig. 4-4-1 Tools

2. Selection of Measuring Tools(Fig. 4-4-2)

Outline dimensions are measured with a vernier caliper. Depth dimensions are measured with a depth vernier caliper. The holes are measured with an inside micrometer.

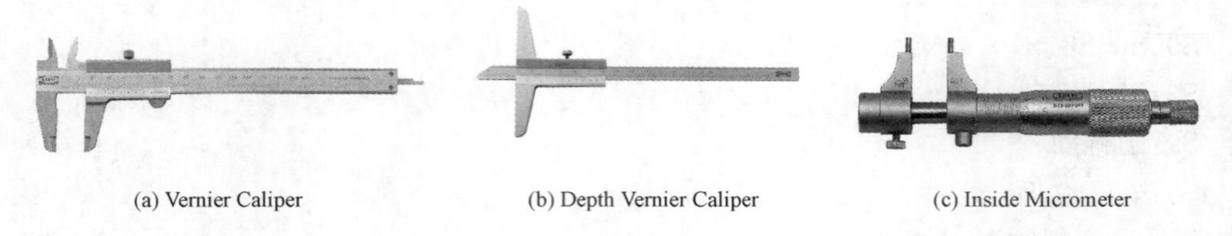

(a) Vernier Caliper　　(b) Depth Vernier Caliper　　(c) Inside Micrometer

Fig. 4-4-2 Measuring Tools

3. Selection of Cutters(Fig. 4-4-3)

The face milling cutter is used for the milling of the upper surface. The keyway cutter is used for machining contour. Holes are drilled with the center drill, twist drill and reamer.

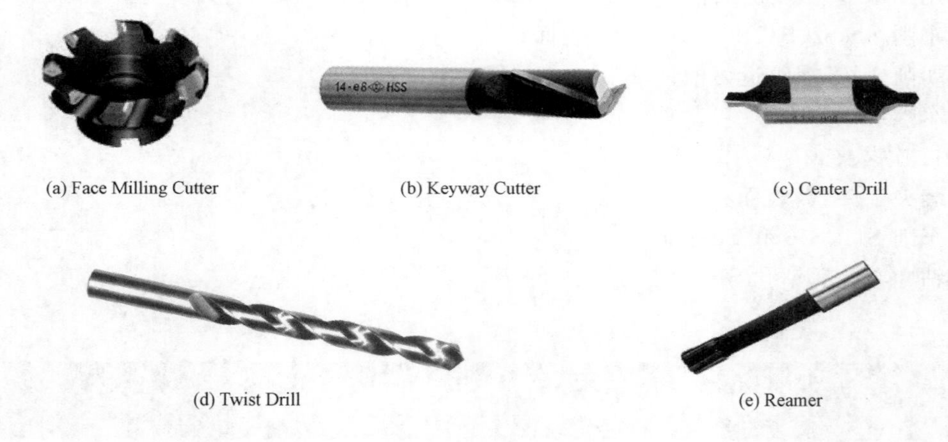

(a) Face Milling Cutter　　(b) Keyway Cutter　　(c) Center Drill

(d) Twist Drill　　(e) Reamer

Fig. 4-4-3 Cutters

4. Selection of Milling Mode

There are two ways of peripheral milling, which are the climb milling and the conventional milling. For the climb milling, the rotational direction of the milling cutter is the same with the direction of cutting feed. If the rotational direction of the milling cutter is opposite to the direction of cutting feed, it is the conventional milling. In case of spindle running forward, if the tool path is clockwise around the outer contour of the workpiece, it is the climb milling; if the tool path is counterclockwise, it is the conventional milling. On the contrary, when milling inner contour of the workpiece, if the tool path is clockwise, it is the conventional milling; if the tool path is counterclockwise, it is the climb milling, as shown in Fig. 4-4-4.

Because the tools are mounted on the spindle of the vertical machining center, elastic bending deformation will produce in the cutting. When the tool is in the climb milling, it will produce inadequate cutting. In the conventional milling, it will produce overcutting. This case becomes more obvious when the tool diameter is smaller and the shank extends longer. So the climb milling is usually selected for roughing. Both the climb milling and the conventional milling can be used for semi-finishing or finishing.

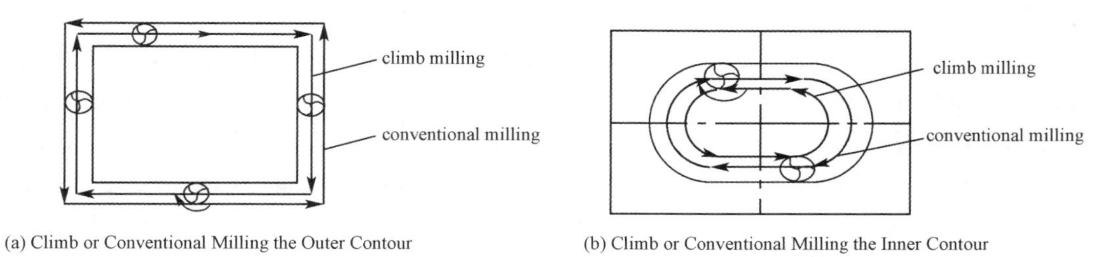

(a) Climb or Conventional Milling the Outer Contour (b) Climb or Conventional Milling the Inner Contour

Fig. 4-4-4 Climb Milling and Conventional Milling

5. An Example of Machining Process

(1) Part Drawing

Part Drawing is shown in Fig 4-4-5.

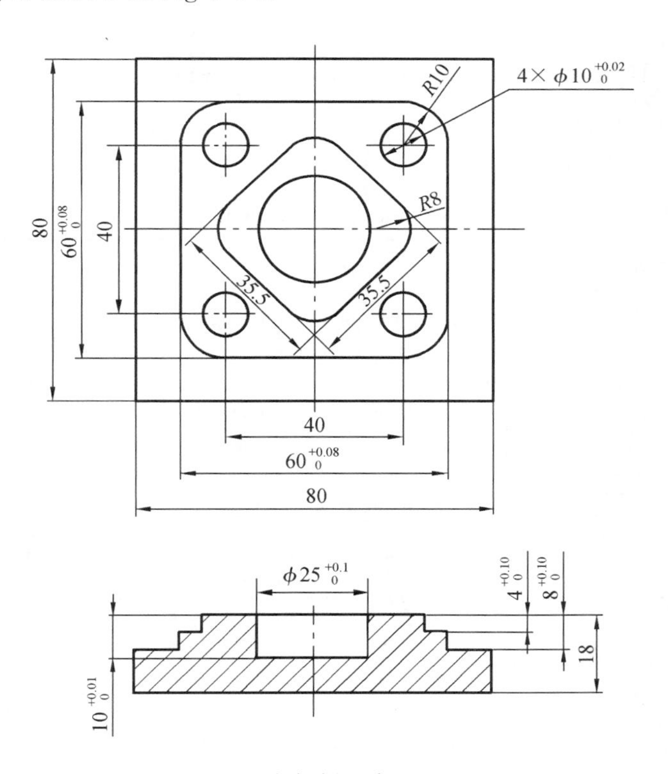

Technical Requirements

1. Unspecified dimensional tolerances should meet the requirements of GB/T 1184.
2. Break sharp corners.
3. After processing, no burr is allowed on the part.

Fig. 4-4-5 Part Drawing

(2) Machining Process Route

Main milling sites of the workpiece are outlines of the steps, flat bottom holes, drillings and reamings. For machining the outer contour, the method of first roughing and then finishing is adopted. Only finishing program will be written in a single production. By modifying tool radius, allowances for finishing are set during roughing machining. For mass production, roughing and finishing programs should be prepared in order to improve production efficiency. According to the precision requirements of holes, it needs drilling and reaming after the surface machining. If the first step is drilling, reaming and then machining on the surface, aluminum scrap will fall into the machined hole, leading to scratches on the hole's wall. Specific process routes are as follows:

- Rough mill and finish mill the upper surface of work blank, rough milling margin under rough conditions controlled by programs, leaving finish milling margin 0.5 mm;
- Rough mill and finish mill the outer contour and $\phi25^{+0.1}_{0}$ mm hole with $\phi16$ keyway milling cutter;
- Drill $4\times\phi10^{+0.02}_{0}$ mm center holes with a A2 center drill;
- Drill $4\times\phi10^{+0.02}_{0}$ mm bottom holes with a $\phi9.7$ twist drill;
- Ream $4\times\phi10^{+0.02}_{0}$ mm holes with a $\phi10$H8 machine reamer.

(3) Selecting Cutting Dosages

Reference cutting dosages are in the Chart 4-4-1.

Chart 4-4-1 **Reference Cutting Dosages**

tool No.	tool specification	process contents	$v_f/(\text{mm}\cdot\text{min}^{-1})$	$n/(\text{r}\cdot\text{min}^{-1})$
T1	$\phi60$ face milling cutter	rough milling and finish milling the upper surface of work blank	100/80	500/800
T2	$\phi16$ keyway milling cutter	rough milling and finish milling the outer contour and inner contour	100	800/1 200
T3	A2 center drill	drill center holes	100	1 000
T4	$\phi9.7$ twist drill	drill $4\times\phi10^{+0.02}_{0}$ mm bottom holes	100	800
T5	$\phi10$H8 machine reamer	ream $4\times\phi10^{+0.02}_{0}$ mm holes	100	1 200

(4) Operating Precautions

- When installing the machine vice, the fixed jaw needs to be calibrated.
- When installing the workpiece, it needs to be placed in the middle of the jaw.
- When the workpiece is installed in the jaw, parallel clamps are needed under the workpiece. When necessary, dial test indicator alignment is used to align with the upper surface of the workpiece in order to keep it horizontal.
- Fluid should be fully filled during machining.

Translation

任务 4 数控铣削加工工艺基础

1. 工具选择(图 4-4-1)

工件装夹在机用虎钳上，机用虎钳用百分表校正。X、Y 方向用寻边器对刀，Z 方向用 Z 轴设定器对刀。

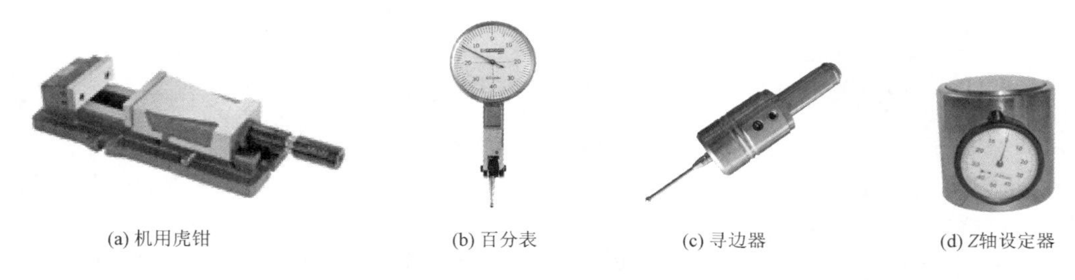

(a) 机用虎钳　(b) 百分表　(c) 寻边器　(d) Z轴设定器

图 4-4-1 工具

2. 量具选择(图 4-4-2)

轮廓尺寸用游标卡尺测量，深度尺寸用深度游标卡尺测量，内孔用内径千分尺测量。

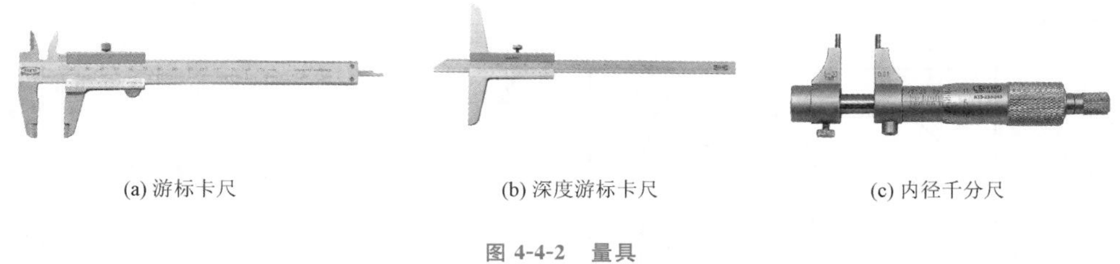

(a) 游标卡尺　(b) 深度游标卡尺　(c) 内径千分尺

图 4-4-2 量具

3. 刀具选择(图 4-4-3)

上表面铣削用面铣刀，轮廓加工用键槽铣刀，孔加工用中心钻、麻花钻及铰刀。

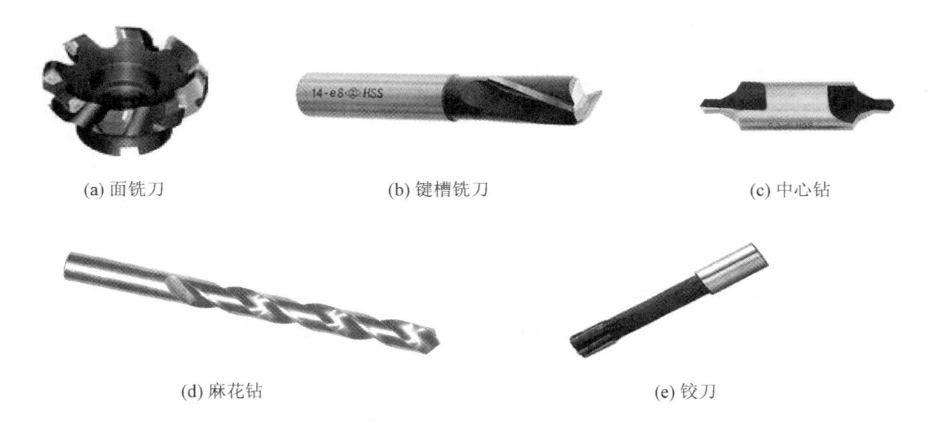

(a) 面铣刀　(b) 键槽铣刀　(c) 中心钻

(d) 麻花钻　(e) 铰刀

图 4-4-3 刀具

4. 铣削方式选择

圆周铣削有顺铣和逆铣两种方式。对于顺铣，铣刀的旋转方向与切削进给方向相同。如果铣刀的旋转方向与切削进给方向相反，则为逆铣。如果主轴正转，则绕工件外轮廓顺时针走刀为顺铣，逆时针走刀为逆铣。相反，当铣削工件内轮廓时，顺时针走刀为逆铣，逆时针走刀为顺铣，如 4-4-4 所示。

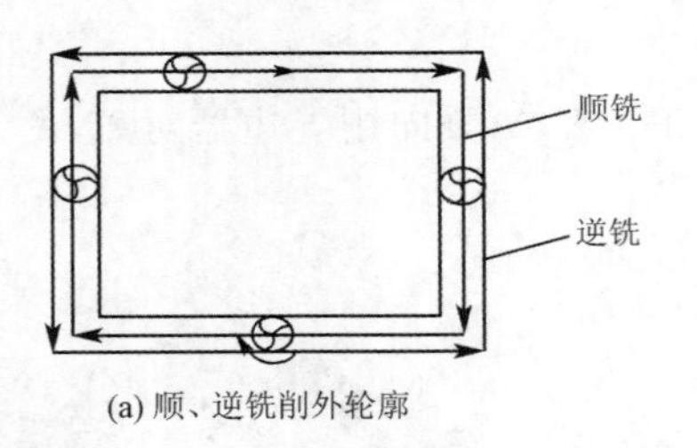

(a) 顺、逆铣削外轮廓

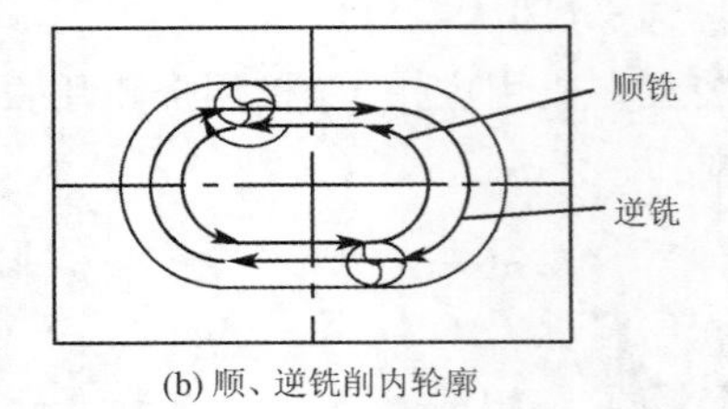

(b) 顺、逆铣削内轮廓

图 4-4-4 顺铣与逆铣

由于刀具装在立式加工中心的主轴上，故切削加工时刀具会产生弹性弯曲变形。当刀具顺铣时，会产生欠切；逆铣时，会产生过切。当刀具直径较小且刀杆伸出较长时，这种情况会更加明显。所以，粗加工时一般采用顺铣；半精加工或精加工既可以采用顺铣，又可以采用逆铣。

5. 铣削加工工艺分析实例

(1)零件图

零件图如图表 4-4-5 所示。

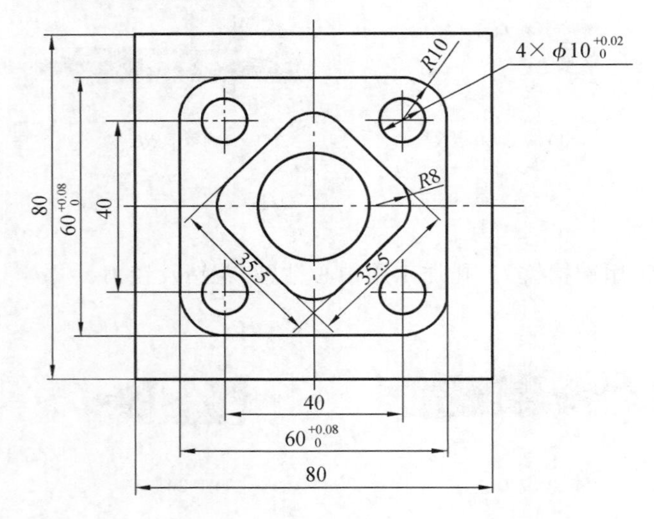

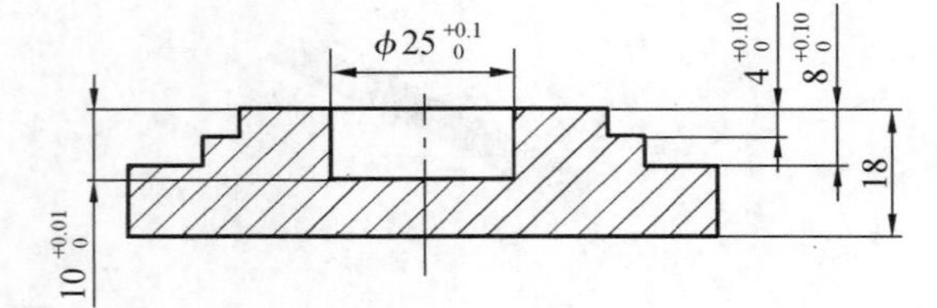

技术要求

1.未注尺寸公差应符合GB/T 1184的要求。
2.锐角倒钝。
3.加工后零件不允许有毛刺。

图 4-4-5 零件图

(2)加工工艺路线

该零件以铣削台阶外轮廓、平底孔、钻孔及铰孔为主。零件外轮廓的加工采用先粗后精的方法。在单件生产中可以只编写精加工程序,通过修改刀具半径补偿完成工件的粗加工。对于大批量生产,应编写粗、精加工程序,以提高生产率。根据被加工孔的精度要求,在表面加工结束后需进行钻孔和铰孔加工。若先钻、铰孔后再加工表面,则铝屑会掉入已加工孔内,使已铰好的孔壁被划伤。具体工艺路线如下:

- 粗、精铣毛坯上表面。粗铣余量根据毛坯情况由程序控制,留精铣余量 0.5 mm;
- 用 ϕ16 键槽铣刀粗、精铣外轮廓和 $\phi25^{+0.1}_{0}$ mm 孔;
- 用 A2 中心钻钻 $4\times\phi10^{+0.02}_{0}$ mm 中心孔;
- 用 ϕ9.7 麻花钻钻 $4\times\phi10^{+0.02}_{0}$ mm 底孔;
- 用 ϕ10H8 机用铰刀铰 $4\times\phi10^{+0.02}_{0}$ mm 孔。

(3)选择切削用量

切削用量参考值见表 4-4-1。

表 4-4-1　　切削用量参考值

刀具号	刀具规格	工序内容	$v_f/(mm\cdot min^{-1})$	$n/(r\cdot min^{-1})$
T1	ϕ60 面铣刀	粗、精铣毛坯上表面	100/80	500/800
T2	ϕ16 键槽铣刀	粗、精铣外轮廓和内轮廓	100	800/1 200
T3	A2 中心钻	钻中心孔	100	1 000
T4	ϕ9.7 麻花钻	钻 $4\times\phi10^{+0.02}_{0}$ mm 底孔	100	800
T5	ϕ10H8 机用铰刀	铰 $4\times\phi10^{+0.02}_{0}$ mm 孔	100	1 200

(4)操作注意事项

- 安装机用虎钳时,要对固定钳口进行校正。
- 安装工件时,要将工件放在钳口的中间部位。
- 在钳口上安装工件时,工件下面要垫平行垫铁。必要时用百分表找正工件上表面,使其保持水平。
- 加工过程中应充分加注切削液。

Task 5 Foundation of Measuring Technology

1. Tolerance

There are two types of tolerances: dimensional tolerance and geometric tolerance.

(1) Dimensional Tolerance

Dimensional tolerance can be symmetrical, such as 99±0.1 (shown in

Fig. 4-3-4), or asymmetrical, such as $\phi 40^{-0.025}_{-0.041}$ (shown in Fig. 4-3-4), in which the dimension 40 is nominal size. The upper limit deviation is −0. 025, and the lower limit deviation is −0. 041. Therefore, the upper limit size is 39. 975, and the lower limit size is 39. 959.

It is often desirable to specify the largest possible tolerance while maintaining proper function. Smaller tolerances are more difficult and costly to achieve.

(2) Geometric Tolerance

Geometric tolerance is used to specify the features of shape and position, such as straightness, flatness, circularity, cylindricity, perpendicularity, etc.. Geometric tolerance characteristic symbols are shown in Chart 4-5-1.

Chart 4-5-1 Geometric Tolerance Characteristic Symbols

Type	Symbol	Type	Symbol
straightness	—	angularity	∠
flatness	▱	parallelism	//
circularity(roundness)	○	position	⌖
cylindricity	⌭	concentricity	◎
profile of line	⌒	symmetry	⌯
profile of surface	⌓	circular run out	↗
perpendicularity	⊥	total run out	⌰

For example, in Fig. 4-3-4, there is a geometric tolerance requirement, which is concentricity [◎|0.03|A].

2. Surface Roughness

Surface roughness is a significant factor to measure its surface quality, such as surface roughness √Ra 1.6 in Fig. 4-3-4. It has great influence on the parts' mating, wear resistance, corrosion resistance, leak proof and appearance. Therefore, desired surface roughness requirement for machining should be specified appropriately on the part drawing.

3. Commonly Used Measuring Tools

(1) Vernier Caliper

● Components of Vernier Caliper

The vernier caliper is a precision instrument that can be used to measure internal and external distances extremely accurately. The examples shown are a manually operated vernier caliper (Fig. 4-5-1) and a digital vernier caliper (Fig. 4-5-2). It usually consists of internal jaws, external jaws, locking screw and depth measuring blade, etc..

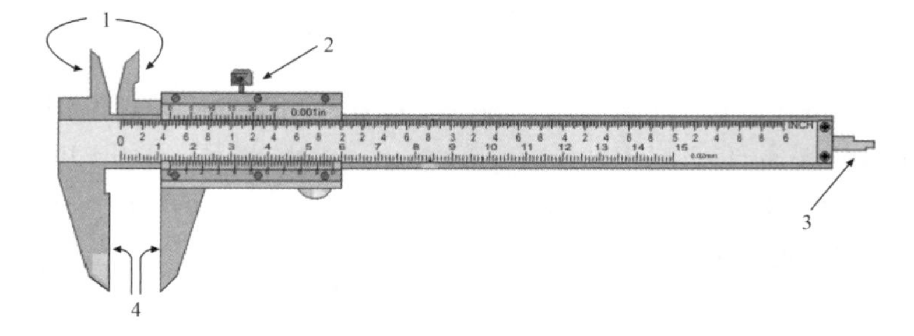

Fig. 4-5-1 Manually Operated Vernier Caliper

1—internal jaws; 2—locking screw; 3—depth measuring blade; 4—external jaws

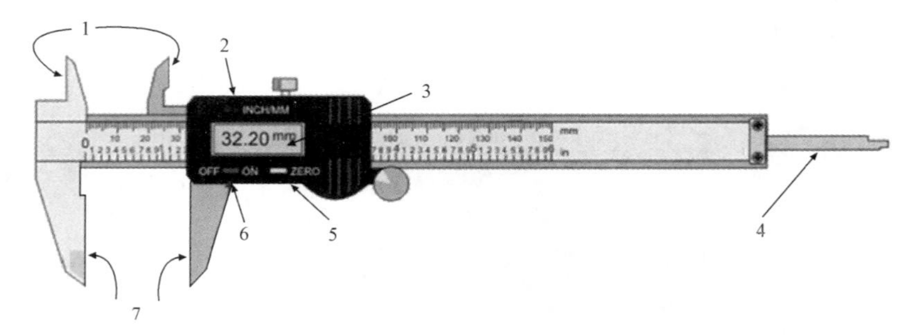

Fig. 4-5-2 Digital Vernier Caliper

1—internal jaws; 2—metric/inch button; 3—LED display; 4—depth measuring blade; 5—zero setting; 6—power on/off; 7—external jaws

◎ Reading of Vernier Caliper

• Read the Whole Number on the Main Scale

The main scale is read by taking note of where the zero mark on the vernier scale falls on the main scale. This is the whole number that should be noted.

• Read the Decimal on the Vernier Scale

The decimal is read from the vernier scale. This number is taken as the line on the vernier scale that aligns with any line on the main scale.

• Add Them All Up

Add the whole number and the decimal together as the final measuring result.

For example, in Fig. 4-5-3, it shows that there are 13 whole divisions before the zero on the vernier scale. Therefore, the whole number is 13.

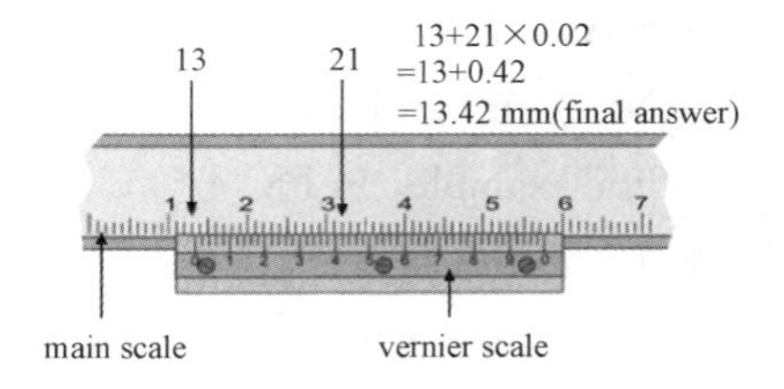

Fig. 4-5-3 Reading of Vernier Caliper

We can also see that the 21st division on the vernier scale lines up exactly with a division on the main scale. This 21 is multiplied by 0.02 giving 0.42 as the answer (each division on the vernier scale is equivalent to 0.02 mm).

◎ Usage of Vernier Caliper

When taking a measurement, you should hold the vernier caliper in the right hand and the workpiece in the left hand. The workpiece to be measured is placed between the external jaws. Then the locking screw is tightened so that the jaws do not move apart.

When measuring the workpiece, the measured surface of jaws must be parallel or vertical to the workpiece surface. The correct reading is making the slight line perpendicular to the vernier caliper. Otherwise the measurement results are not accurate.

(2) Outside Micrometer

Components of Outside Micrometer

Outside micrometer is usually called micrometer for short. It is an even more precise measuring instrument than vernier caliper. Fig. 4-5-4 shows a common outside micrometer, whose range is 0～25 mm. It usually consists of sleeve, thimble, anvil, spindle and ratchet, etc..

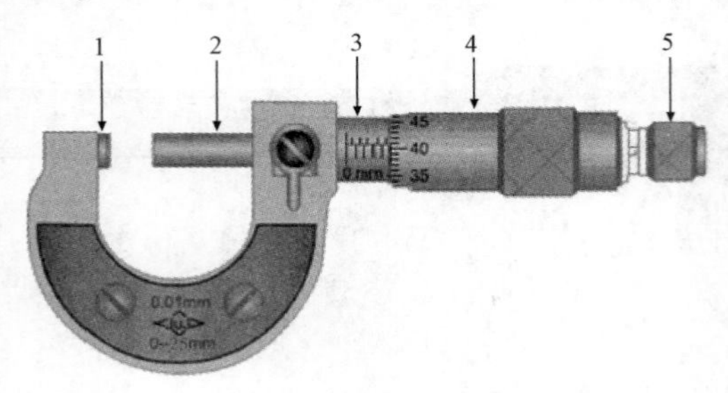

Fig. 4-5-4 Outside Micrometer

1—anvil; 2—spindle; 3—sleeve; 4—thimble; 5—ratchet

Reading of Outside Micrometer

- Read on the Sleeve

To read a micrometer, you place the workpiece to be measured between the anvil and the spindle, then spin the ratchet until the spindle meets the workpiece evenly. Then you can read the markings. Take the thimble's end surface as the directrix, you can read the markings on the sleeve.

- Read on the Thimble

Take the center line on the sleeve as the directrix, read the markings on the thimble.

- Add Them All Up

Now just add all the numbers together to determine the dimension of your workpiece.

For example, in Fig. 4-5-5, we can see that the left of the thimble's end surface has the 8.5 exposed on the sleeve. The next thing is to find the numbers on the thimble which coincides with the center line of the sleeve. In this example, the number 38 is showing on the thimble. Each of those marks is worth 0.01, so we can add another 0.38 to our sum. Finally we add 8.5 and 0.38 together to give the final measurement of 8.88.

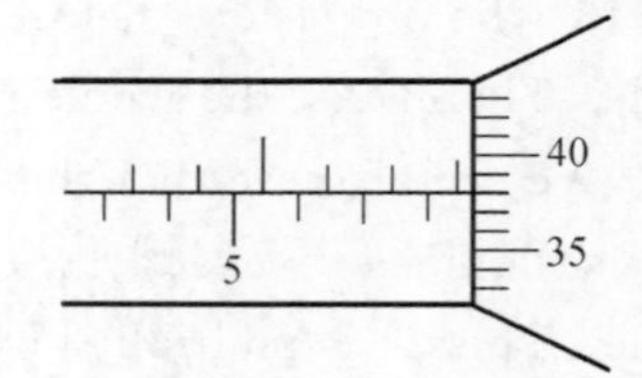

Fig. 4-5-5 Reading of Outside Micrometer

Usage of Outside Micrometer

The proper way to hold the micrometer when taking a measurement is:

- Close the spindle gently against the anvil and note if the zero line on the thimble coincides with the zero on the sleeve.
- Before taking a measurement, the workpiece surface should be cleaned.
- Hold the workpiece in the left hand and the micrometer in the right hand.
- Place the workpiece between the anvil and spindle, then spin the ratchet until you hear three clicks to verify both the anvil and spindle are touching the workpiece evenly.

(3) Thread Gauge

◎ Types of Thread Gauge

Thread gauges include thread plug gauge (Fig. 4-5-6) and thread ring gauge (Fig. 4-5-7).

Fig. 4-5-6 Thread Plug Gauge

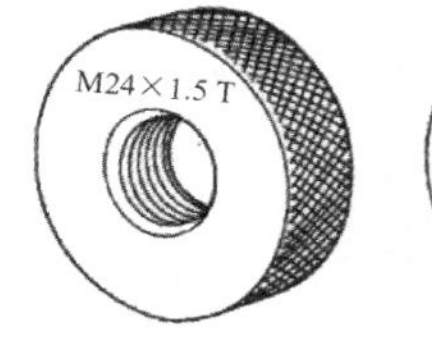

Fig. 4-5-7 Thread Ring Gauge

The thread plug gauge is used to check internal thread. The thread plug gauge is the double-ended go/no-go instrument and similar to the straight hole plug gauge. The gauge ends have different lengths for easy identification, with the long end being the go end and the short end being the no-go end.

The thread ring gauge is used to check external thread. The go and the no-go gauges are separate rings. Generally, the surface of the gauge is stamped with character "T" or "Z" to distinguish the go side from the no-go side.

◎ Usage of Thread Gauge

Using the thread gauge is pretty simple. If the go end of the thread plug gauge goes into the thread hole and the no-go end does not, then the product is usually considered within the tolerance. It is the same for use of the thread ring gauge.

4. Coordinate Measuring Machine (CMM)

A coordinate measuring machine is a device for measuring the physical geometrical characteristics of an object. This machine can be manually controlled by an operator or be computer-controlled. Measurement is defined by a probe attached to the third moving axis of this machine. The probe may be mechanical, optical or other devices.

A coordinate measuring machine is comprised of four main components: the main itself, the measuring probe, the controlling system for computer and the measuring software.

Based on the arrangement of the components moveable in the axis directions, coordinate measuring machines can be divided into four basic types: column type, gantry type, cantilever type and bridge type (Fig. 4-5-8). Coordinate measuring machines of the bridge type are the most widely used. They are used for measuring tasks even when high precision is required.

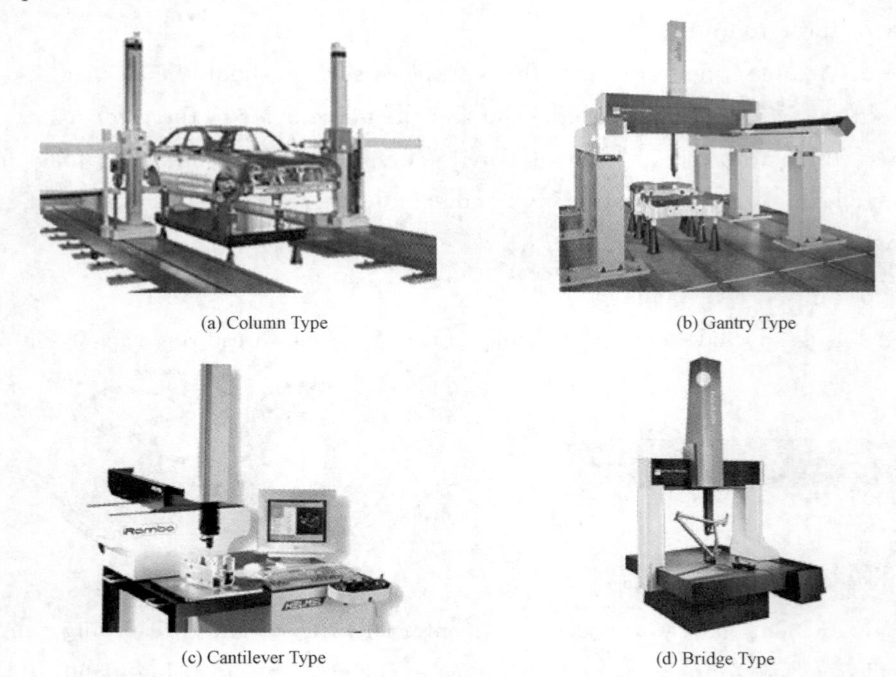

(a) Column Type　(b) Gantry Type

(c) Cantilever Type　(d) Bridge Type

Fig. 4-5-8　Types of CMM

Translation

任务 5　测量技术基础

1. 公差

公差分为两种：尺寸公差和几何公差。

(1)尺寸公差

尺寸公差可以是对称的，如图 4-3-4 中的 99 ± 0.1；或者是非对称的，如图 4-3-4 中的 $\phi40_{-0.041}^{-0.025}$，其中尺寸 40 是公称尺寸，上极限偏差为 -0.025，下极限偏差为 -0.041，因此，上极限尺寸为 39.975，下极限尺寸为 39.959。

通常在保证适当功能的同时，希望指定尽可能大的公差值。获得小的公差非常困难而且成本高。

(2)几何公差

几何公差用来指定形状和位置特征，比如直线度、平面度、圆度、圆柱度、垂直度等。几何公差特征符号见表 4-5-1。

表 4-5-1 几何公差特征符号

种类	符号	种类	符号
直线度	⏤	倾斜度	∠
平面度	▱	平行度	//
圆度	○	位置度	⌖
圆柱度	⌭	同轴度	◎
线轮廓度	⌒	对称度	⌯
面轮廓度	⌓	圆跳动	↗
垂直度	⊥	全跳动	⌰

例如，在图 4-3-4 中有几何公差要求，即同轴度要求 ◎|0.03|A 。

2. 表面粗糙度

表面粗糙度是衡量零件表面质量的重要因素，如图 4-3-4 中的表面粗糙度 $\sqrt{Ra\ 1.6}$ 。它对零件的配合、耐磨性、抗腐蚀性、密封性以及外观都有很大的影响。因此，加工后想要达到的表面粗糙度要求应在零件图上准确提出。

3. 常用量具

(1)游标卡尺

◎游标卡尺的组成

游标卡尺是一种能非常准确地测量内、外部距离的精密测量仪器。图 4-5-1 是普通游标卡尺，图 4-5-2 是数显游标卡尺。游标卡尺通常由内测量爪、外测量爪、紧固螺钉和深度尺等组成。

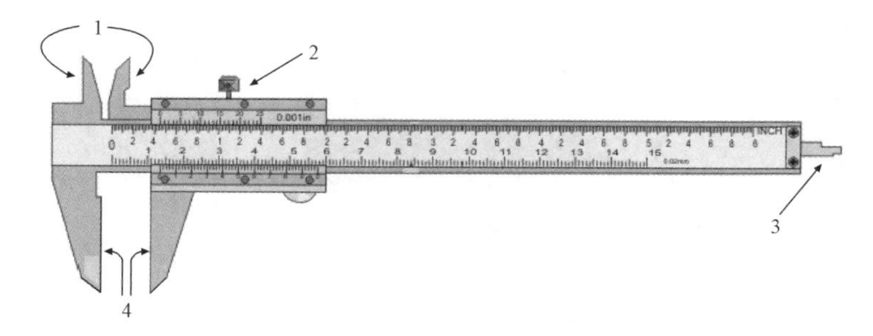

图 4-5-1 普通游标卡尺

1—内测量爪；2—紧固螺钉；3—深度尺；4—外测量爪

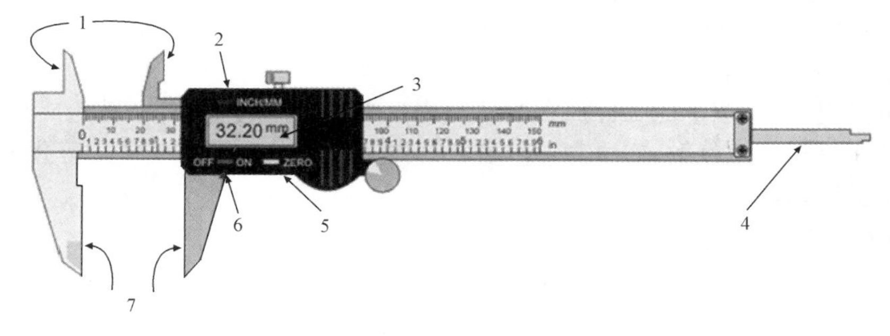

图 4-5-2 数显游标卡尺

1—内测量爪；2—公/英制按钮；3—液晶显示屏；4—深度尺；5—零点设定；6—电源开/关；7—外测量爪

◎游标卡尺的读数

• 在主尺上读整数

主尺上整数的读数方法：读出游标尺零线左面尺身上的数。值得注意的是，这个数是整数。

• 在游标尺上读小数

从游标尺上读出小数部分。小数部分是通过找出游标尺上与主尺上的哪条刻线对齐而得到的。

• 两者相加

将整数部分和小数部分相加，得到最终测量结果。

例如图 4-5-3 中，游标尺零线左面尺身上有 13 条刻线，因此整数部分为 13。

我们还可以看到，游标尺上第 21 条刻线恰好与主尺上的一条线对齐。将 21 乘以 0.02，得到 0.42 为小数部分（游标尺上的每一个刻度值为 0.02 mm）。

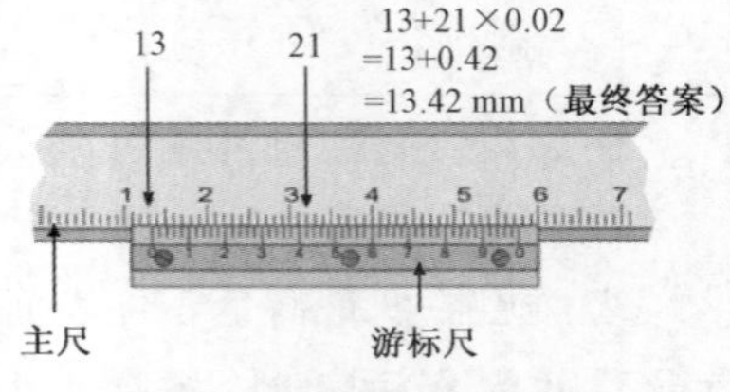

图 4-5-3　游标卡尺的读数

◎游标卡尺的使用

测量时，右手拿住游标卡尺，左手拿住工件。被测工件位于两个外测量爪之间。然后锁紧紧固螺钉，使卡爪不分离。

测量工件时，卡爪的测量面必须与工件的表面平行或垂直。读数时视线要垂直于尺面，否则测量结果不准确。

（2）外径千分尺

◎外径千分尺的组成

外径千分尺常简称为千分尺，它是一种比游标卡尺更精密的测量仪器。图 4-5-4 所示为一种常见的量程为 0～25 mm 的外径千分尺，它通常由固定套筒、微分筒、测砧、测微螺杆和棘轮旋柄等组成。

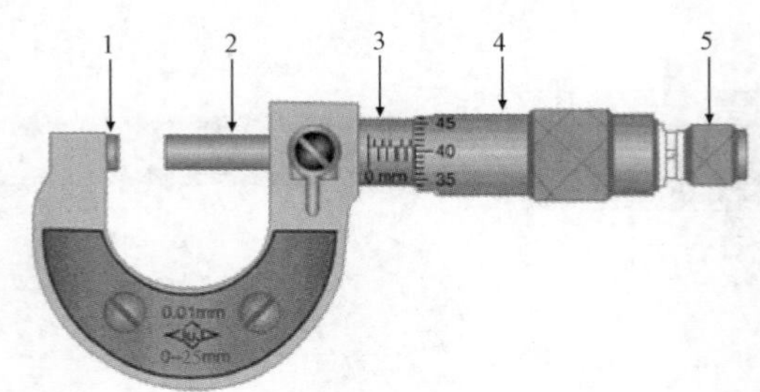

图 4-5-4　外径千分尺寸

1—测砧；2—测微螺杆；3—固定套筒；4—微分筒；5—棘轮旋柄

◎外径千分尺的读数

• 在固定套筒上读数

读千分尺时，把被测量的工件放在测砧和测微螺杆之间，然后转动棘轮旋柄，直到测微螺杆和被测工件很好地接触，之后就可以进行读数了。以微分筒的端面为基准线，可以读出固定套筒上的数值。

• 在微分筒上读数

以固定套筒上的中心线作为基准线，读出微分筒上的数值。

- 两者相加

把这两部分的数值相加，就是工件的尺寸。

例如图 4-5-5 中，微分筒端面左边的 8.5 显示在固定套筒上，然后找到微分筒上与固定套筒对齐的刻线数。在本例中数值 38 显示在微分筒上。每一条刻线的值为 0.01，于是我们可以将数值 0.38 加到总数中。最后把 8.5 和 0.38 相加，即得出最终的测量结果 8.88。

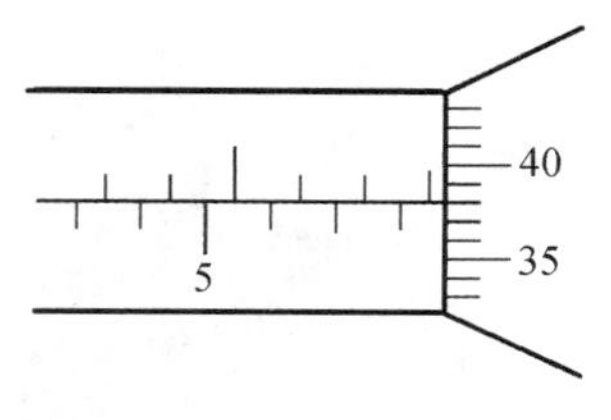

图 4-5-5　外径千分尺的读数

◎外径千分尺的使用

测量时，正确使用外径千分尺的方法如下：

- 使测微螺杆与测砧轻轻地接触，注意微分筒上的零线是否与固定套筒上的零线对齐。
- 测量前，应将工件表面擦干净。
- 左手握住工件，右手拿住千分尺。
- 将工件放在测砧与测微螺杆之间，然后旋转棘轮旋柄，直到听到三声响后，核实测砧和测微螺杆与工件接触是否良好。

(3)螺纹量规

◎螺纹量规的种类

螺纹量规包括螺纹塞规(图 4-5-6)和螺纹环规(图 4-5-7)。

图 4-5-6　螺纹塞规

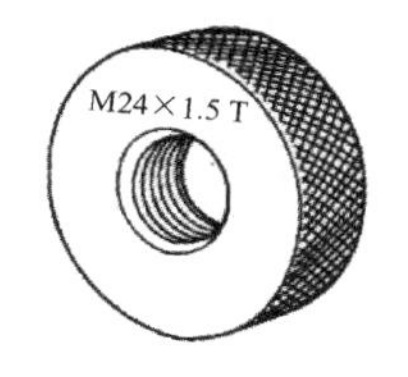

图 4-5-7　螺纹环规

螺纹塞规用于检测内螺纹。螺纹塞规与直孔用塞规相似，具有通端和止端。量规的两端长度不等，很容易区分，长的一端为通规，短的一端为止规。

螺纹环规用于检测外螺纹。螺纹环规的通端和止端是分开的。通常在量规的表面印有字母“T”和“Z”来区分通规和止规。

◎螺纹量规的使用

螺纹量规的使用非常简单。如果螺纹塞规的通端能通过螺纹孔而止端不能通过，我们通常就认为该产品在公差极限范围内。螺纹环规的使用也是一样的。

4. 三坐标测量仪(CMM)

三坐标测量仪是一种能够测量物体几何特征的仪器。这种仪器可以由操作者操作或由计算机控制。其测量是通过与仪器第三根动轴连在一起的探头进行的。探头可以是机械的，光学的或者其他装置。

三坐标测量仪由四部分构成：仪器本身、测量探头、计算机控制系统和测量软件。

根据移动部件沿坐标轴方向的布局，可将三坐标测量仪分为四种基本类型：立柱式、龙门式、悬臂式和桥式(图 4-5-8)。桥式三坐标测量仪是最常用的类型，用于测量精度要求高的产品。

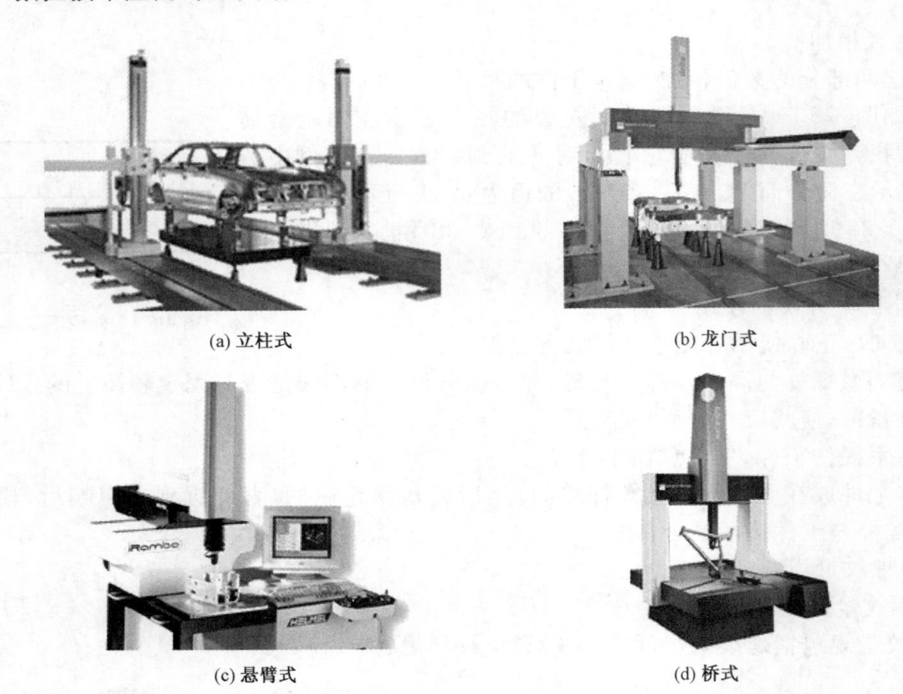

(a) 立柱式 (b) 龙门式

(c) 悬臂式 (d) 桥式

图 4-5-8 三坐标测量仪的类型

学习启示

科学家冲破一切阻碍回国，一干就是几十年，无怨无悔，就是抱着一个强国的梦想。爱国是做人的第一要义，要把聪明才智贡献给自己的祖国，让祖国强大，让人民富裕。

Module 5
Advanced Manufacturing Technology

Task 1 CNC Nontraditional Machining

1. CNC Sinker EDM

(1) Basic Principle of Sinker EDM

Sinker electrical discharge machining (sinker EDM), also known as Die-EDM, is that the electrode and the workpiece are separately connected to the two poles of pulse power supply. The pulse power supply generates an electrical potential between the two parts. As the electrode approaches the workpiece, dielectric breakdown occurs in the fluid forming a plasma channel and a small spark jump. The metal is removed by a series of discrete electrical discharges (sparks) causing localized temperature that is high enough to melt or vaporize the metal. As the base metal is eroded and the spark gap subsequently increases, the electrode is lowered automatically by the machine so that the process can continue uninterruptedly. The constant pulse discharge will replicate the tool electrode shape on the workpiece to carry out the shaped machining, as shown in Fig. 5-1-1.

(2) CNC Sinker EDM Machine

A CNC sinker EDM machine is composed of mechanical unit, pulse power, servo mechanism, CNC system, dielectric fluid filter, electrode and so on. Fig. 5-1-2 illustrates the appearance of CNC sinker EDM machine.

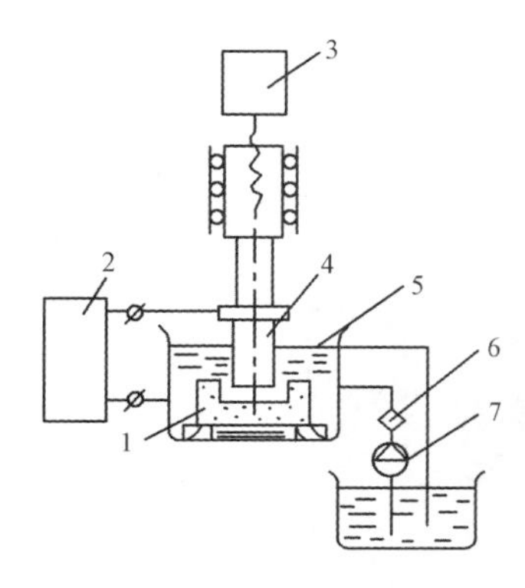

Fig. 5-1-1 Basic Principle of Sinker EDM

1—workpiece; 2—pulse power; 3—servo mechanism; 4—electrode; 5—dielectric fluid; 6—dielectric fluid filter; 7—pump

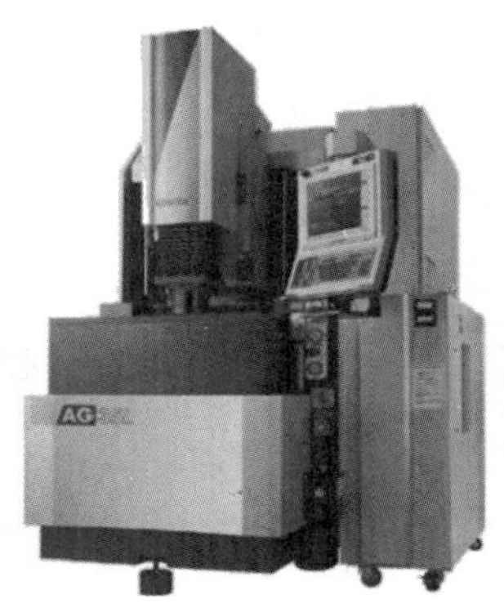

Fig. 5-1-2 Appearance of CNC Sinker EDM Machine

2. CNC WEDM

(1)Basic Principle of WEDM

Wire cut electrical discharge machining(WEDM), is also known as wire cutting. The basic principle is shown in Fig. 5-1-3. WEDM is an electrical discharge machining process with a continuously moving conductive wire as the tool electrode. The mechanism of metal removal in wire electrical discharge machining involves the complex erosion effect of electric sparks generated by a pulsating direct current power supply between two closely spaced electrodes. The energy with high density erodes material from both the wire and the workpiece by local melting and vaporizing. Because the new wire keeps feeding to the machining area, and the material of the workpiece is removed by the moving of wire electrode. Eventually, a cutting shape is formed on the workpiece by the programmed path of wire electrode.

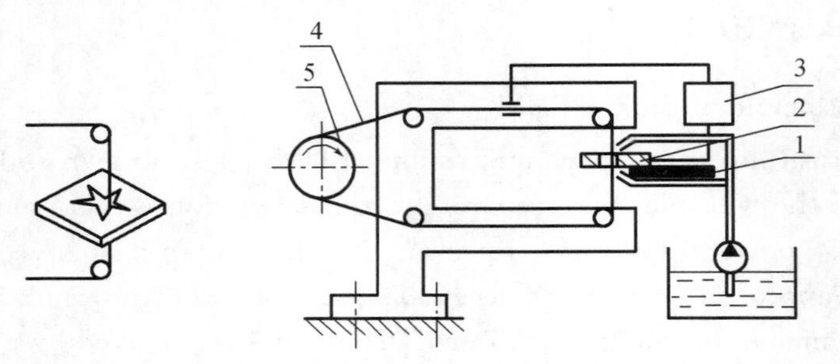

Fig. 5-1-3 Basic Principle of WEDM

1—insulation base plate; 2—workpiece; 3—pulse power;
4—wire electrode; 5—wire spool

(2)CNC WEDM Machine

CNC WEDM machine can be classified into the following two kinds according to the wire electrode's movement speed:

◎ Low-speed CNC WEDM Machine

Wire electrode moves in one-way low-speed, and the speed of wire is 0.2 m/s. In order to maintain the stability of molybdenum wire, to enhance the quality of workpiece surface, we can reduce the wire speed, thus it is named low-speed CNC WEDM. Low-speed CNC WEDM can obtain better surface roughness and accuracy than high-speed CNC WEDM.

The low-speed CNC WEDM machine is shown in Fig. 5-1-4.

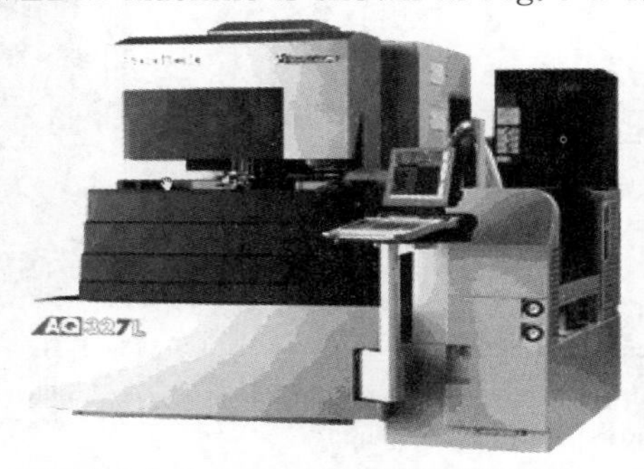

Fig. 5-1-4 Low-speed CNC WEDM Machine

◎ High-speed CNC WEDM Machine

High-speed CNC WEDM machine is the special local product in China, which is widely used in mold industry. Its wire electrode moves back and forth in a high speed. General wire speed is 8～10 m/s. It is much cheaper than a low-speed CNC WEDM machine. On the quality, a high-speed CNC WEDM machine can satisfy most of the customers' demands well. So it is getting more and more popular worldwide.

The high-speed CNC WEDM machine is shown in Fig. 5-1-5.

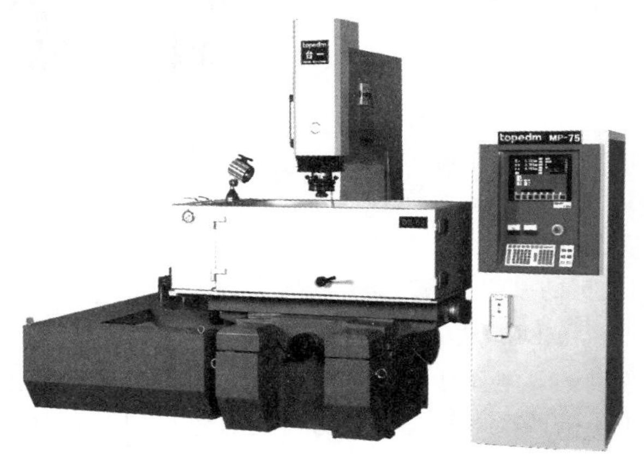

Fig. 5-1-5 High-speed CNC WEDM Machine

Translation

模块 5 先进制造技术

任务 1 数控特种加工

1. 数控电火花成型加工

(1)电火花成型加工基本原理

电火花成型加工(sinker EDM)也被称为模具电火花加工，是将电极和工件分别连接到脉冲电源的两极。脉冲电源在电极和工件之间产生一个电场。当电极靠近工件时，电解液被击穿形成一个放电通道，并且产生一个小的火花放电。一系列不间断的火花放电所产生的局部高温使金属熔化或气化，从而去除材料。当底部的材料被去除后，放电间隙随之增大，机床通过自动降低电极来保持不间断的放电过程。这种不间断的脉冲放电将电极的形状复制到工件上，以实现成型加工，如图 5-1-1 所示。

(2)数控电火花成型加工机床

数控电火花成型加工机床由机床本体、脉冲电源、伺服机构、数控系统、电解液过滤器及电极等部分组成。图 5-1-2 所示为数控电火花成型加工机床的外观。

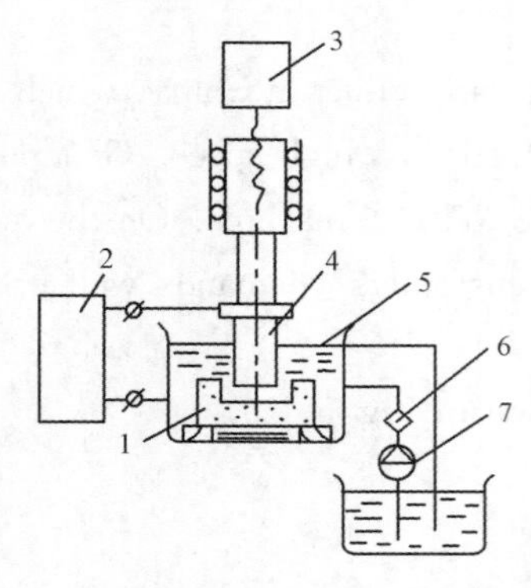

图 5-1-1 电火花成型加工基本原理

1—工件；2—脉冲电源；3—伺服机构；4—电极；5—电解液；6—电解液过滤器；7—泵

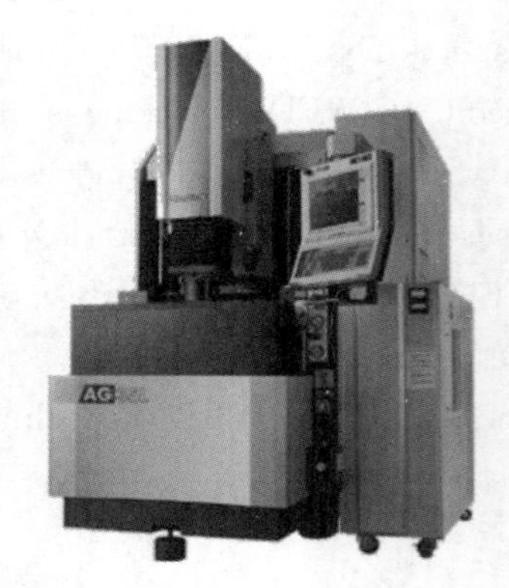
图 5-1-2 数控电火花成型加工机床的外观

2. 数控电火花线切割加工

(1)电火花线切割加工的基本原理

电火花线切割加工(WEDM)也被称为线切割，其基本原理如图 5-1-3所示。电火花线切割加工是利用不断移动的导电丝作为工具电极的放电加工过程。电火花线切割加工去除金属的机理是在间距很小的两个电极之间接脉冲直流电源，从而产生火花放电的复合侵蚀作用。高密度能量通过局部熔融和气化，来蚀除电极丝和工件上的材料。由于新的电极丝不断进给到加工区域，工件上的材料不断被移动的电极丝去除，故最终电极丝按照编程路径切割出需要的工件形状。

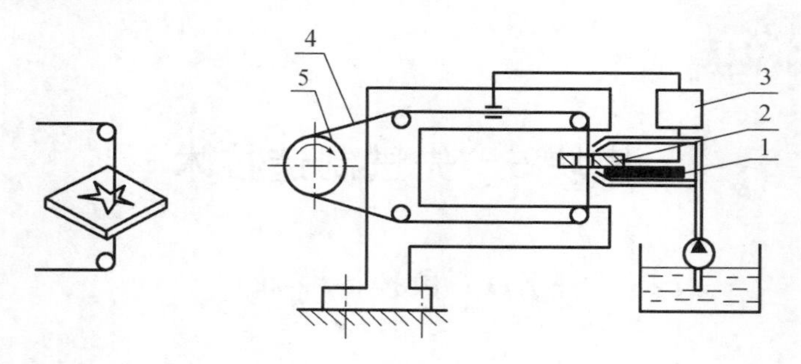

图 5-1-3 电火花线切割加工的基本原理

1—绝缘底板；2—工件；3—脉冲电源；4—电极丝；5—丝筒

(2)数控电火花线切割机床

根据电极丝的移动速度，可以将数控电火花线切割机床分为两类：

●慢走丝数控电火花线切割机床

电极丝单向慢速移动，移动速度为 0.2 m/s。为了保持钼丝的稳定性、提高工件表面加工质量而降低走丝速度，因此它被命名为慢走丝数控电火花线切割加工。慢走丝数控电火花线切割加工可以获得比快走丝数控电火花线切割加工更好的表面粗糙度和精度。

图 5-1-4 所示为慢走丝数控电火花线切割机床。

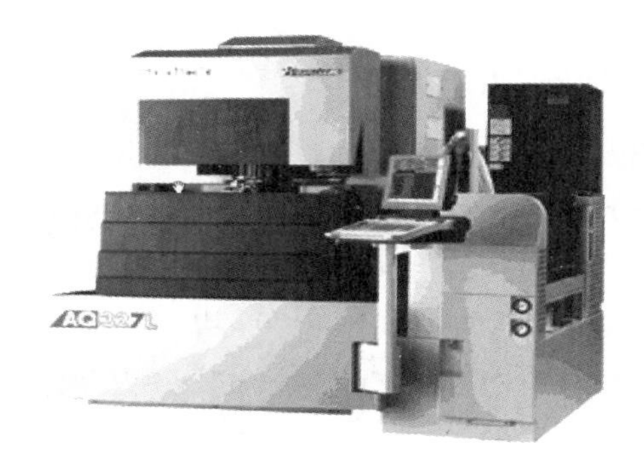

图 5-1-4 慢走丝数控电火花线切割机床

◎快走丝数控电火花线切割机床

快走丝数控电火花线切割机床是中国独创的，广泛应用于模具行业中。它的电极丝是双向往复高速运行的。一般电极丝速度为 8～10 m/s。它比慢走丝数控电火花线切割机床更加便宜。在加工质量上，快走丝数控电火花线切割机床能满足大多数客户的要求，因此它在世界上也变得越来越流行。

图 5-1-5 所示为快走丝数控电火花线切割机床。

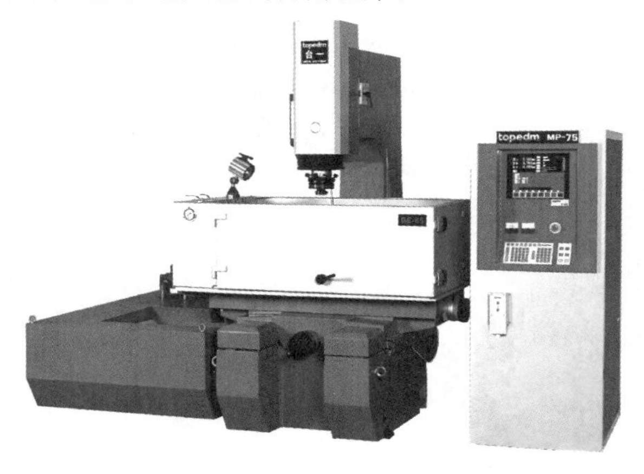

图 5-1-5 快走丝数控电火花线切割机床

Task 2 Robot

1. Robot Overview

课文音频

A robot is a programmable, multi-functional device designed to move material, parts, tools, or special devices through variable programmed motions for the performance of a variety of tasks. Robots are especially suited for places where jobs are extremely risky, tedious and repetitive, demanding high accuracy and places that do great harm to humans.

2. Classifications of Industrial Robots

Robots have gradually shifted from manufacturing to non-manufacturing and service industries. The industrial robots are the most mature and widely used category of robots. Service industries include cleaning, refueling, ambulance, rescue, disaster relief, medical treatment and so on. According to the application, the industrial robots can be divided into welding robots, stacking robots, sealing robots, cutting robots, loading and unloading robots and so on.

(1) Welding Robot (Fig. 5-2-1)

The welding robot is a kind of industrial robot in welding production realm to replace a welder to engage the welding task. Some of the welding robots are specially designed for some welding methods. Most welding robots are actually general-purpose industrial robots equipped with some welding tools.

There are two main ways of the welding robot's applications, namely, spot welding and electric arc welding. The application of industrial robots in the field of welding is the earliest starting from the resistance spot welding of the automobile assembly line. The arc welding robot is not only used in the automobile manufacturing industry, but also in other manufacturing industries involving arc welding, such as shipbuilding, locomotives, boilers, heavy machinery, etc..

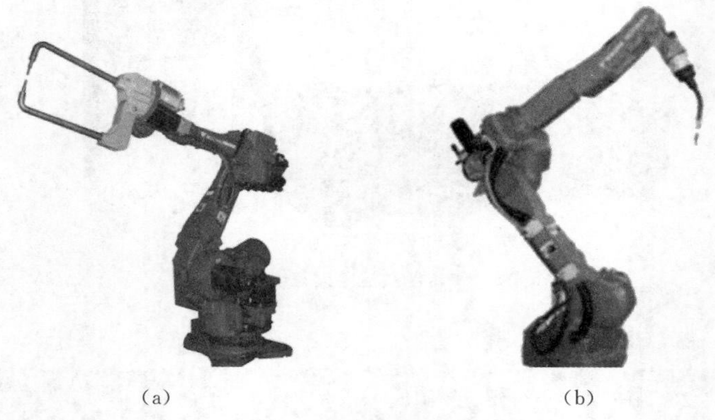

(a) (b)

Fig. 5-2-1 Welding Robot

(2) Stacking Robot (Fig. 5-2-2)

The stacking robot mainly involves the operations during storing and handling in delivery process and warehouse of the factory. Stacking robots could stack a large number of various products on the pallet according to the order in a short time.

(3) Sealing Robot (Fig. 5-2-3)

The sealing robot is an applicator installed at the front end of the manipulator, for doing sealants, filler, solder coating. The sealing robot must coat the sealing parts continuously and uniformly.

Fig. 5-2-2 Stacking Robot

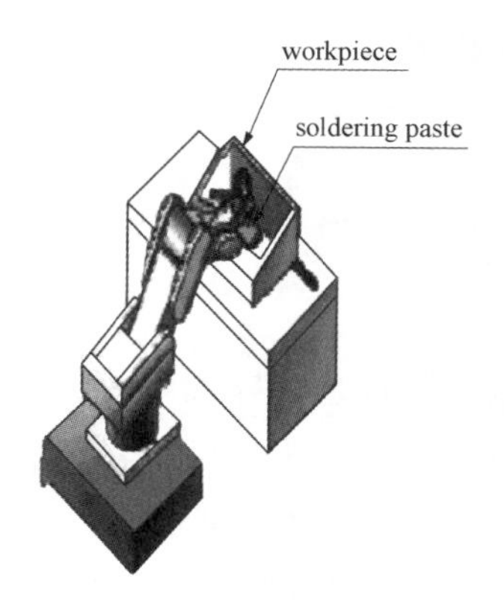

Fig. 5-2-3 Sealing Robot

(4) Cutting Robot (Fig. 5-2-4)

The cutting robot is the cutting tools (pliers, etc.) installed at the front end of the manipulator to do the cutting.

(5) Loading and Unloading Robot (Fig. 5-2-5)

The loading and unloading robot is applied to load raw workpieces on the CNC machine and unload the finished workpieces after processing.

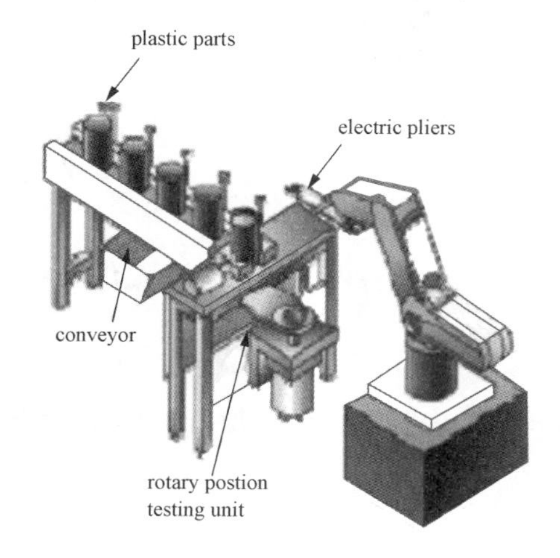

Fig. 5-2-4 Cutting Robot

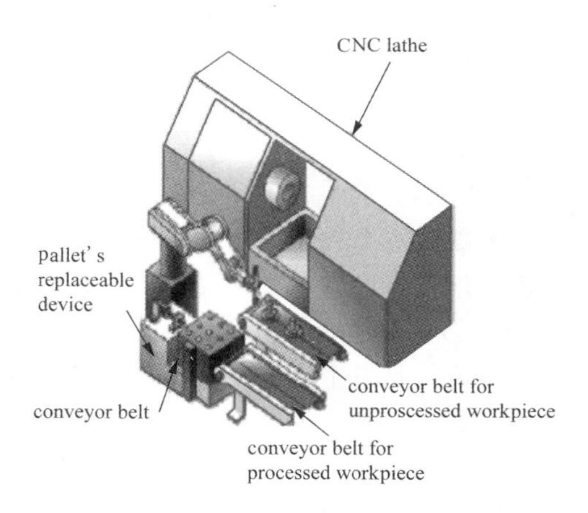

Fig. 5-2-5 Loading and Unloading Robot

任务2 机器人

1. 机器人概述

机器人是一种通过不同的程序动作来完成多种工作的可编程、多功能的装置，其设计目的是用来搬运材料、零部件、工具或特殊装置。机器人特别适用于极度危险的场所、枯燥而重复的工作、要求高精度的工作和对人类有很大危害的地方。

2. 工业机器人的分类

机器人已经从制造业逐渐转向了非制造业和服务行业。工业机器人是最成熟、应用最广泛的一类机器人。服务行业包括清洁、加油、救护、抢险、救灾、医疗等。按照应用领域，工业机器人可以分为焊接机器人、码堆机器人、密封机器人、切割机器人和装卸机器人等。

(1)焊接机器人(图 5-2-1)

焊接机器人是在焊接生产领域代替焊工从事焊接任务的工业机器人。焊接机器人中有的是为某种焊接方式专门设计的，而大多数的焊接机器人其实就是通用的工业机器人装上某种焊接工具而构成的。

焊接机器人的应用主要有两种，即点焊和电弧焊。工业机器人在焊接领域的应用最早是从汽车装配生产线上的电阻点焊开始的。电弧焊机器人不仅用于汽车制造业，还用于涉及电弧焊的其他制造业，如造船、机车车辆、锅炉、重型机械等。

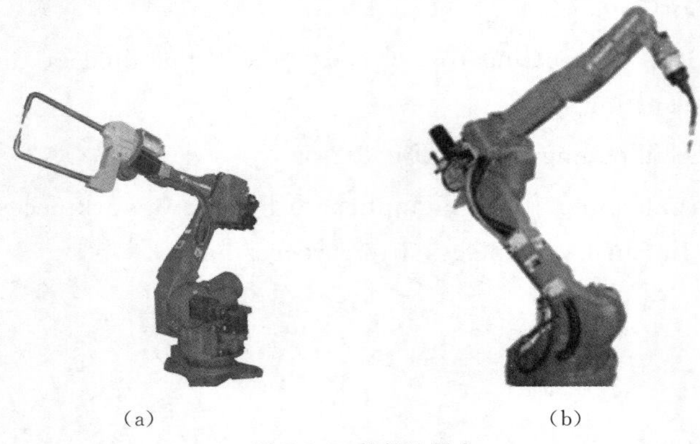

(a)　　(b)

图 5-2-1　焊接机器人

(2)码堆机器人(图 5-2-2)

码堆机器人主要在产品出厂工序和仓库的储存保管时进行作业。码堆机器人能够在短时间内按照订单将大量各类产品堆积在托盘上。

(3)密封机器人(图 5-2-3)

密封机器人是一个安装在机械手前端的涂敷头，用来进行密封剂、填料、焊料涂敷等作业。密封机器人必须对密封部位进行连续、均匀的涂敷。

图 5-2-2　码堆机器人

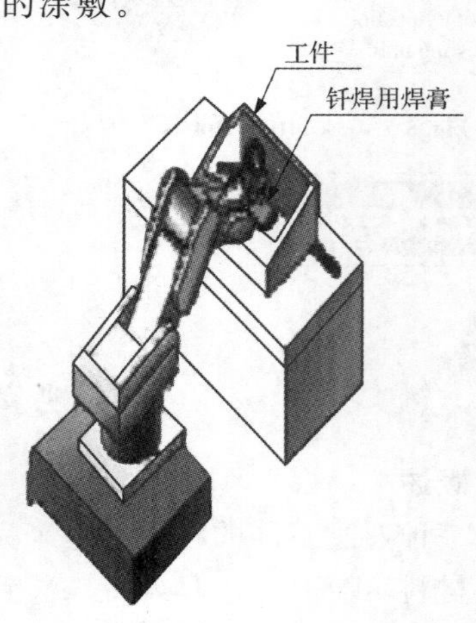

图 5-2-3　密封机器人

(4)切割机器人(图 5-2-4)

切割机器人是安装在机械手前端的切割工具(剪钳等),用来进行切割作业。

(5)装卸机器人(图 5-2-5)

装卸机器人用于在数控机床上装夹未加工的工件,并且将加工后的工件取下。

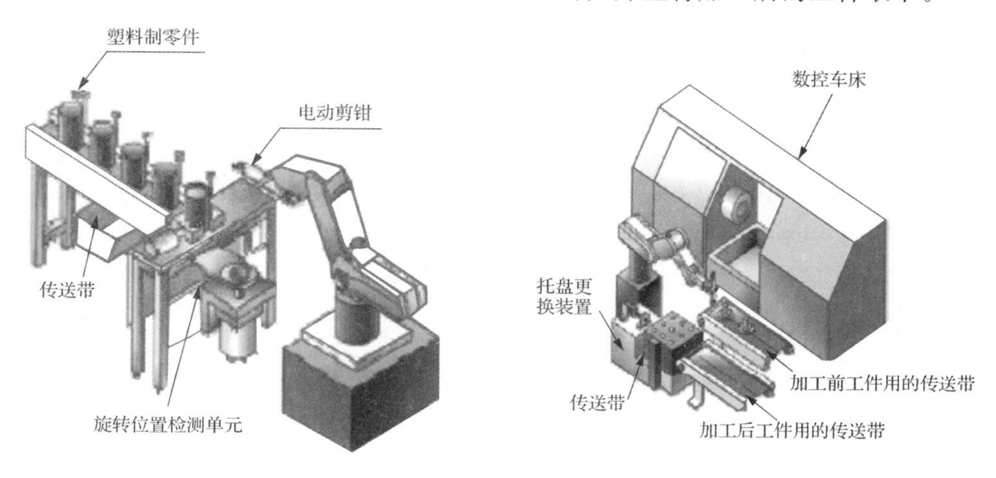

图 5-2-4 切割机器人

图 5-2-5 装卸机器人

Task 3 3D Printing

1. 3D Printing Overview

课文音频

3D printing refers to the use of three-dimensional inkjet printing technology, through the combination of layered processing and superposition forming method, to increase the material to generate 3D entities layer by layer, to the same 3D real object digital manufacturing as by laser molding and other 3D model manufacturing technologies.

3D printing process (Fig. 5-3-1) is: first modeling through the computer modeling software, and then building the three-dimensional model "partition" into layers, that is slices, to guide the printer layer by layer. The standard file format for collaboration between the design software and the printer is the STL file format.

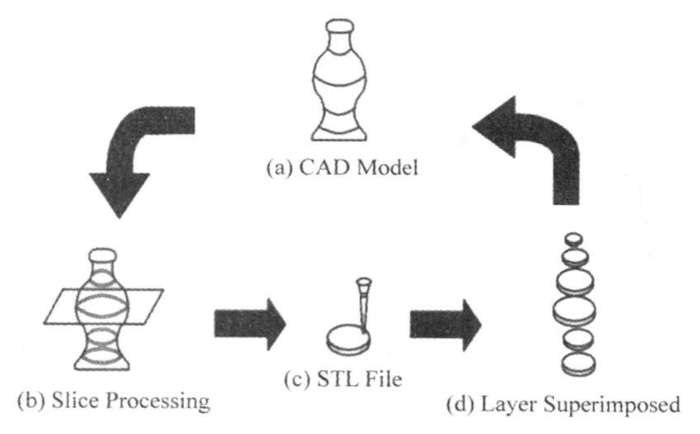

Fig. 5-3-1 3D Printing Process

The 3D printer, based on the three-dimensional printing technology, first sends a certain amount of raw material powder by the storage barrel, which is pushed into a thin layer in the processing platform by the drum, and then the print head sprays a special adhesive in the area that needs shaping. At this point, the powder encountered by the adhesive will quickly cure the bond, and the powder not encountered by the adhesive is still in loose state. Each time the layer is finished, the processing platform will automatically drop a little. According to the results of computer slices, it continues to cycle until the physical completion. After the completion, as long as the removal of loose outer powder the required three-dimensional objects can be obtained.

The 3D printer contains different printed materials such as metal, ceramic, plastic, and sand. After the printer and the computer are connected, the printed material can be cascaded through the computer to eventually turn the blueprint on the computer into an object.

2. Classification of 3D Printing

There are many different techniques for 3D printing. At present, the most popular 3D printing consists of fused deposition modeling and selective laser sintering.

(1) Fused Deposition Modeling (FDM)

FDM is the abbreviated form of Fused Deposition Modeling, which is melt deposition. Popularity is the use of high temperature melting the material into liquid, which, after extrusion curing through the print head, is finally arranged in the three-dimensional space to form three-dimensional objects. This manufacturing technique is generally applicable to thermoplastics. The working principle of the 3D printer based on the FDM technique is shown in Fig. 5-3-2.

(2) Selective Laser Sintering (SLS)

A layer of powder material is first coated with a powder roll, then heated to a temperature below the sintering point of the powder by means of a constant temperature facility in the printing apparatus, and then the laser beam irradiates on the powder layer to increase its temperature above the melting point, so that the irradiated powder is sintered and bonded to the portion which has been formed below. When a layer is finished after sintering, the printing platform drops a layer of height, the powder delivery system paves the new powder material for the printing platform, and then controls the laser beam to re-irradiate the powder for sintering, forming the cycle, stacking until the entire 3D object printing job is finished (Fig. 5-3-3).

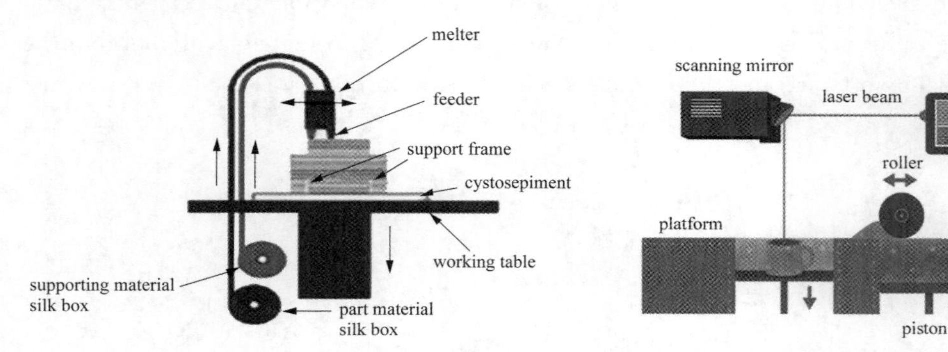

Fig. 5-3-2 Working Principle of Fused Deposition Modeling

Fig. 5-3-3 Working Principle of Selective Laser Sintering

3. Applications of 3D Printing

At present, 3D printing, supporting a wide range of materials, can be widely used in jewelry, footwear, industrial design, construction, automotive, aerospace, dental, medical and even food. In different areas, according to the needs of application targets, the materials used are resin, nylon, gypsum, ABS, polycarbonate(PC) or ingredients and so on. The rapid prototyping technology of the 3D printer makes it unique in the market and has great potential in production applications.

4. Development Trend of 3D Printing

3D printing technology continues to develop. With a substantial reduction of its cost, it has been developed from the small space to the mainstream market. The momentum of its development cannot be stopped. It has become a widespread concern in the community and rapidly rising new civilian market areas. In the next 10 years, you can complete the product design blueprint on the computer, then gently press the "print" button, and the 3D printer will be able to print the designed model. Some foundry companies are now developing selective laser sintering 3D printers and their applications, trying to reduce the delivery time of complex castings from 3 months to 10 days. Engine manufacturers have reduced the large six-cylinder diesel engine cylinder head sand core development cycle from the past 5 months to 1 week through 3D printing technology. The advantage of 3D printing technology is to expand the designer's imagination.

3D printing will eliminate the traditional production line, shorten the production cycle, greatly reduce the production of waste, and the required amount of raw materials will be reduced several times the original amount. 3D printing not only saves costs and improves the production accuracy, but also will make up for the disadvantages of traditional manufacturing and rise rapidly in the civilian market, thus opening a new era of manufacturing and bringing new opportunities and hope for the printing industry.

任务3　3D打印

1. 3D打印概述

3D打印是指采用三维喷墨打印技术，通过分层加工与叠加成型相结合的方法，逐层打印增加材料来生成3D实体，达到与激光成型等其他3D模型制造技术相同的3D真实物体数字制造技术。

3D打印过程(图5-3-1)是：先通过计算机建模软件建模，再将建成的三维模型"分区"成逐层的截面，即切片，从而指导打印机逐层打印。设计软件和打印机之间协作的标准文件格式是 STL 文件格式。

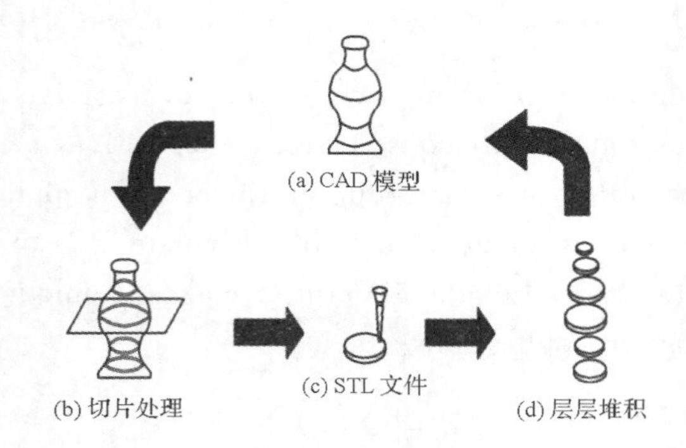

图 5-3-1 3D 打印过程

基于三维打印技术的 3D 打印机，先由存储桶送出一定量的原材料粉末，粉末在加工平台上被滚筒推成薄薄一层，接着打印头在需要成型的区域喷出一种特殊的黏结剂。此时，遇到黏结剂的粉末会迅速固化黏结，而没有遇到黏结剂的粉末则仍保持松散状态。每喷完一层，加工平台就会自动下降一点，根据电脑切片的结果不断循环，直到实物完成。完成之后，只要扫除松散的外层粉末，便可获得所需要的三维实体。

3D 打印机内装有金属、陶瓷、塑料、砂等不同的打印材料。当打印机与计算机连接后，通过计算机可以把打印材料一层层地叠加起来，最终把计算机上的蓝图变成实体。

2. 3D 打印的分类

3D 打印存在着许多不同的技术。目前，比较流行的 3D 打印主要包括熔融沉积成型和选择性激光烧结。

(1)熔融沉积成型(FDM)

FDM 是 Fused Deposition Modeling 的缩写形式，即熔融沉积成型。通俗来讲就是利用高温将材料融化成液态，通过打印头挤出后固化，最后在三维空间上排列形成三维实体。这种制造技术普遍适用于热塑性塑料。基于熔融沉积制造技术的 3D 打印机的工作原理如图 5-3-2 所示。

(2)选择性激光烧结(SLS)

先用铺粉滚轴铺一层粉末材料，通过打印设备里的恒温设施将其加热至恰好低于该粉末烧结点的某一温度，接着激光束在粉层上照射，使被照射的粉末温度升至熔化点之上，进行烧结并与下面已制作成型的部分实现黏结。当一个层面完成烧结之后，打印平台下降一个层厚的高度，铺粉系统为打印平台铺上新的粉末材料，然后控制激光束再次照射进行烧结，如此循环往复，层层叠加，直至完成整个三维实体的打印工作(图 5-3-3)。

3. 3D 打印的应用

目前 3D 打印支持多种材料，可以广泛应用在珠宝首饰、鞋类、工业设计、建筑、汽车、航天、牙科、医疗甚至食品等领域。在不同的领域，根据应用目标的需求，所使用的材料有树脂、尼龙、石膏、ABS、聚碳酸酯(PC)或食材配料等。3D 打印机的快速成型技术使之在市场上独具优势，在生产应用上潜力巨大。

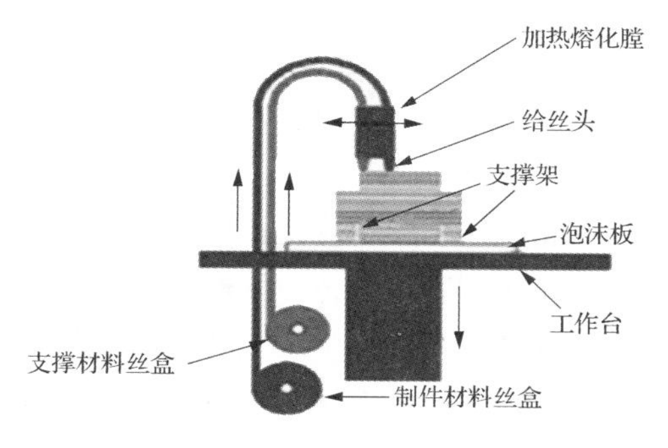

图 5-3-2 熔融沉积成型的工作原理

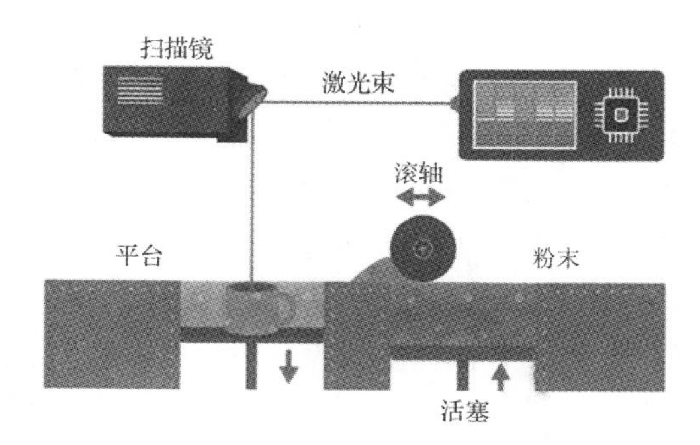

图 5-3-3 选择性激光烧结的工作原理

4. 3D 打印发展趋势

3D 打印技术持续发展，成本的大幅降低使其已经从研发的小众空间向主流市场进军，发展势头不可阻挡，已经成为社会广泛关注、民用市场迅速崛起的新领域。未来 10 年在电脑上完成产品设计蓝图，然后轻轻按下“打印”键，3D 打印机就能打印出所设计的模型。现在一些铸造企业开始研发选择性激光烧结 3D 打印机及其应用，力图将复杂铸件的交货期由 3 个月缩短到 10 天。发动机制造商通过 3D 打印技术，将大型六缸柴油发动机缸盖砂芯的研制周期由过去的 5 个月缩短至 1 周。3D 打印技术的优势在于能拓展设计人员的想象空间。

3D 打印将淘汰传统生产线，缩短制造周期，大大减少生产废料，所需原材料用量将减少到原来的几分之一。3D 打印不仅节约成本，提高制造精度，还将弥补传统制造的不足，并将在民用市场迅速崛起，从而开启制造业的新纪元，为印刷工业带来新的机遇和希望。

Task 4 FM and CIM

1. Flexible Manufacturing(FM)

(1)Concept of Flexible Manufacturing Production Line

A flexible manufacturing production line is a production line consisting of a plurality of adjustable machines(mostly dedicated machines) linked together with automatic delivery devices.

(2)Composition of Flexible Manufacturing Production Line

A flexible manufacturing production line is composed of automatic processing system, logistics system, information system and software system.

(3)Forms of Flexible Manufacturing Production Line

● Flexible Manufacturing Cell(FMC)(Fig. 5-4-1)

FMC, consisting of one or two machining centers, has the functions of different cutters' changing and workpiece loading and unloading, delivering and storing. Besides the CNC device, there is a cell computer to manage the programs and peripheral devices. FMC fits for the small batch production and the workpieces of complicated shapes which need less machining processes but long time processings.

◎ Flexible Manufacturing System(FMS)(Fig. 5-4-2)

FMS consists of two or more machining centers, cleaning and testing devices. It is a better delivering and storage system for the cutters and workpiece. Apart from scheduling computer, it has process control computers and distributed numerical control terminals to form the local networks with multi-stage control system. It's an entire unmanded, software-based manufacturing/assembly line. An FMS consists of four major components: the CNC machines, coordinate measuring machines, workpiece handling and assembly robots, and workpiece/tool transfer device.

Fig. 5-4-1 Flexible Manufacturing Cell(FMC)

Fig. 5-4-2 Flexible Manufacturing System(FMS)

◎ Autonomy Manufacturing Island(AMI)

AMI is a manufacturing system consisting of several CNC machines and common machines based on group technology. It is characterized by combining process equipment with production organization management and manufacturing process. It performs the process design, CNC program management, operation planning and real-time production scheduling by computer. It has a wide range of applications, a low investment and high flexibility.

(4) Advantages of Flexible Manufacturing Production Line

◎ High Utilization

The capacity of one set of machine increases several times than that of the single operation after being put on the flexible production line.

◎ High Quality

The automatic processing system is composed of one or more machines. It is capable of degradation operation in case of fault. Material delivery system also has the ability to detour the malfunctioned machine.

◎ Stable Form

The workpiece is loaded and unloaded once to complete the process during the processing. It has high precision and stable processing form.

◎ Flexible Operation

Some of the testing, loading and maintenance can be completed during the first shift. Unmanned normal production can be achieved during the second and third shift. In an ideal flexible manufacturing production line, the monitoring system can also solve some unexpected problems, like the exchange of the worn tools, clearance of the blocked logistics, etc..

- Great Adaptability

Tools, fixtures, and material delivery devices are adjustable. Layout of the system is reasonable and suitable for changing the equipments to meet the market needs.

2. Computer Integrated Manufacturing(CIM)

(1)Definition of CIM

In modern manufacturing, integration is accomplished by computers. CIM can now be defined as the total integration of all manufacturing elements through the use of computers.

(2)Historical Development of CIM

CIM has evolved through the four distinct stages: manual manufacturing, mechanization/specialization, automation, integration.

(3)Functions and Features of CIM

CIM uses computers to integrate the following functions: workpiece and product design; tool and fixture design; the routes, operations, machines, and tools involved in process planning; monitoring costs, manufacturing time, and machine idle time while maximizing productivity and quality; programming of CNC machines and material handling systems; produce planning; manufacturing; assembly; repairing machine malfunctions and quality issues within manufacturing; controlling quality of system design, parameter design, and tolerance design; inspecting errors and quality issues during product manufacturing; storing and retrieving of raw material, work-in-process inventory, finished goods and equipment.

CIM is helpful to increase productivity, improve product quality, decrease costs of processing and reduce scrap and inventory.

(4)Computer Integrated Manufacturing System(CIMS)

CIMS is the abbreviation of computer integrated manufacturing system. It is produced with the development of CAD and CAM. CIMS describes a new approach of manufacturing, management, and corporate operation. CIMS is a kind of real flexible manufacturing system. It can manufacture various kinds of workpieces and components in batches without changing the structure of the system. CIMS includes software and automation systems needed to complete the entire process. It includes product design, system programming, production cost estimation, product actual manufacturing, instruction input, inventory tracking and actual production cost analysis.

Translation

任务4 柔性制造(FM)与计算机集成制造(CIM)

1. 柔性制造(FM)

(1)柔性制造生产线的概念

柔性制造生产线是把多台可以调整的机床(多为专用机床)连接起来,配以自动运送装置的生产线。

(2)柔性制造生产线的构成

柔性制造生产线由自动加工系统、物流系统、信息系统和软件系统构成。

(3)柔性制造生产线的类型

◎柔性制造单元(FMC)(图 5-4-1)

FMC 通常由 1~2 台加工中心构成,具有不同形式的刀具交换、工件装卸、输送及存储功能。除了机床的数控装置外,还有一个单元计算机来进行程序和外围设备的管理,FMC 适合于小批量生产,加工工序不多而加工时间较长的形状比较复杂的零件。

◎柔性制造系统(FMS)(图 5-4-2)

FMS 由 2 台或 2 台以上的加工中心以及清洗、检测设备组成,具有较完善的刀具和工件的输送和存储系统。除调度管理计算机外,它还配有过程控制计算机和分布式数控终端,形成由多级控制系统组成的局部网络。它是一个基于软件的完全无人化生产/装配线。FMS 由四个主要部分组成:数控机床、坐标测量机、工件处理与装配机器人和工件/刀具传输装置。

图 5-4-1 柔性制造单元(FMC)

图 5-4-2 柔性制造系统(FMS)

◎自主制造岛(AMI)

AMI 自主制造岛是以成组技术为基础,由若干台数控机床和普通机床组成的制造系统。其特点是将工艺装备、生产组织管理和制造过程结合在一起,借助计算机进行工艺设计、数控程序管理、作业计划编制和实时生产调度等。其使用范围广,投资相对较少,柔性较高。

(4)柔性制造生产线的优点

◎利用率高

一组机床编入柔性生产线后,产量比这组机床在分散单机作业时提高数倍。

◎产品质量高

自动加工系统由一台或多台机床组成。发生故障时，它能够降级运转。物料传送系统也有自行绕过故障机床的能力。

◎形式稳定

在加工过程中，装卸零件一次即完成加工。加工精度高，加工形式稳定。

◎运行灵活

有些检验、装夹和维护工作可在第一班完成，第二、第三班可在无人照看的情况下正常生产。在理想的柔性制造生产线中，监控系统还能处理诸如磨损刀具的调换、堵塞物流的疏通等运行过程中不可预料的问题。

◎应变能力强

刀具、夹具及物料运输装置具有可调性，且系统平面布置合理，便于增减设备，满足市场需求。

2. 计算机集成制造(CIM)

(1)CIM 的定义

在现代制造中，集成化是由计算机来实现的，我们可以这样定义 CIM，即运用计算机把所有制造环节完全集成起来。

(2)CIM 的历史变迁

CIM 制造业经历了四个不同的发展阶段：手工制造，机械化/专业化，自动化，集成化。

(3)CIM 的功能与特点

CIM 使用计算机集成下述功能：工件和成品设计；工具和夹具设计；工艺流程所涉及的路线、操作、机床和工具；控制成本、加工时间以及生产率最高和质量最佳时的机床空运行时间；对数控机床和材料处理系统编程；做生产计划；加工；装配；加工过程中维修机床故障和品质问题；控制系统设计、参数设计和公差设计的质量；检验产品加工过程中的错误和品质问题；存储和提取原材料、库存、成品和设备。

CIM 有助于提高产量和产品质量以及降低工艺成本、废料和库存。

(4)计算机集成制造系统(CIMS)

CIMS 是 computer integrated manufacturing system 的缩写。它是随着计算机辅助设计与制造的发展而产生的。CIMS 描述了一种制造、管理和公司运作的新方法。CIMS 是一种真正的柔性制造系统，不需要对系统做结构上的改变，就可以批量地制造各种零部件。CIMS 包括完成全部加工过程所需要的软件和自动化系统，包括产品设计、系统编程、生产成本估算、产品实际制造、指令输入、库存跟踪以及实际生产成本分析环节。

学习启示

了解中国机器人的领军企业和领军人物，努力奋斗，学习工匠精神，增强民族自信心和自豪感，坚定四个自信。

Module 6
Integrated Applications

Task 1 Reading Product Specifications

1. CJX2 Series AC Contactor

(1) Applicable Range

CJX2 series AC contactor(Fig. 6-1-1) is applied to the power system with AC 50 Hz or 60 Hz, rated insulation voltage up to 690～1 000 V and rated working current up to 95 A under AC-3 usage sort with rated working voltage 380 V. It is used for on and off circuit from long distance and to frequently start and control AC motor. It also can be combined with proper thermal relay or electronic protection device to form electromagnetic starter to protect the circuit or AC motor from overload.

Fig. 6-1-1 CJX2 Series AC Contactor

(2) Structural Feature

CJX2 series AC contactor has features of small size, lightweight, long life and small power consumption. It can be derived upon request and can be equipped with accessories such as auxiliary contact group, air delay head and mechanical interlock mechanism, etc. to form delay contactor, mechanical interlock contactor and star triangle starter. It also can be directly connected with thermal relay to form electromagnetic starter.

(3) Main Technical Parameters(Chart 6-1-1)

Chart 6-1-1 Main Technical Parameters

Type	Rated Working Current/A	Rated Working Voltage/V	Rated Insulation Voltage/V	Electrical Life/ 10^4 times	Operating Frequency/ h^{-1}	Coil Power/ (V · A)		Main Load Can Be On
						Pull In	Hold	
CJX2-9	9	380	660	100	1 200	70	8	6 V 10 mA
CJX2-32	32	380	660	80	600	110	11	

2. Linear Encoder for CNC Machine

(1) Advantages of Linear Encoder

The linear encoder(Fig. 6-1-2) measures the position of linear axes without additional mechanical transfer element. The control loop for position control with a linear encoder includes the entire feed mechanics. Transfer errors from the mechanics can be detected by the linear encoder on the slide, and corrected by the control circuit. It can eliminate a number of potential error sources:

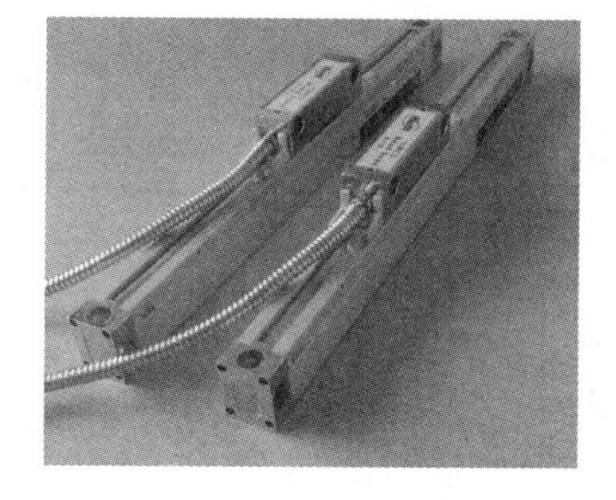

Fig. 6-1-2 Linear Encoder

- Positioning error due to thermal behavior of the ball screw;
- Ball screw backlash;
- Kinematics error through ball screw pitch error.

(2) Mechanical Construction

The linear encoder for numerically controlled machines uses sealed construction. Aluminum housing protects the scale, scanning unit and its guide way from chips, dust and fluids. The elastic lips seal the housing. The scanning unit travels in a low-friction guide within the scale unit(Fig. 6-1-3).

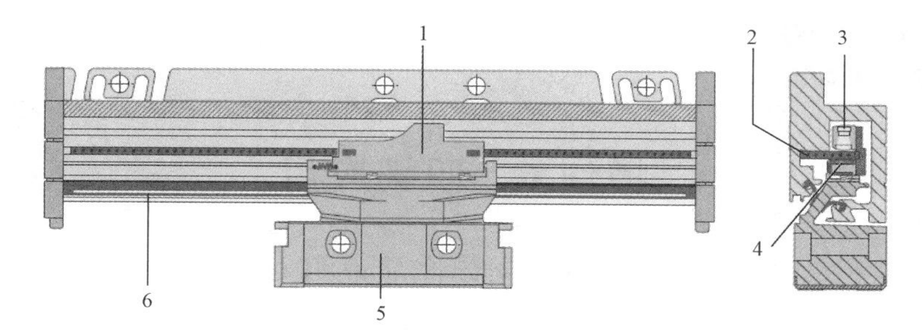

Fig. 6-1-3 Linear Encoder's Mechanical Construction

1—scanning unit; 2—scale; 3—light source; 4—photocells; 5—mounting block; 6—sealing lips

(3) Mounting Guidelines

It is surprisingly simple to mount the sealed linear encoder. You only need to align the scale unit at several points along the machine guide way. Limit surface or limit pin can also be used for this. The shipping brace has already set the proper gap between the scale unit and the scanning unit, as well as the lateral tolerance. The mounting method is shown in Fig. 6-1-4.

The linear encoder should be fastened to a machined surface over its entire length, especially for high-dynamic requirements. The encoder is mounted with the sealing lips directing downward to keep away from splashing water(Fig. 6-1-5).

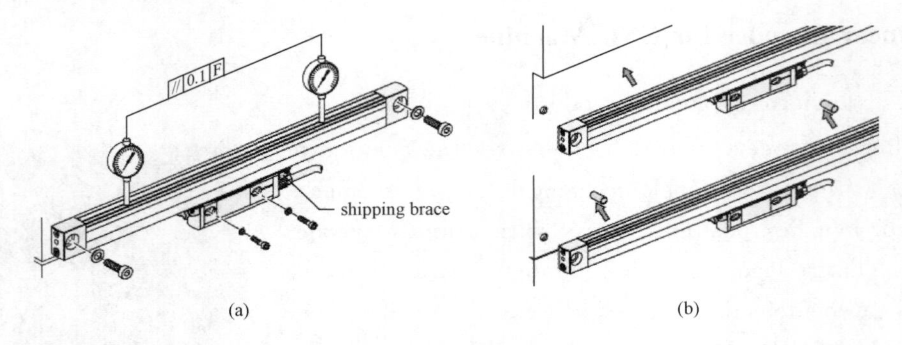

(a) (b)

Fig. 6-1-4 Mounting Linear Encoder

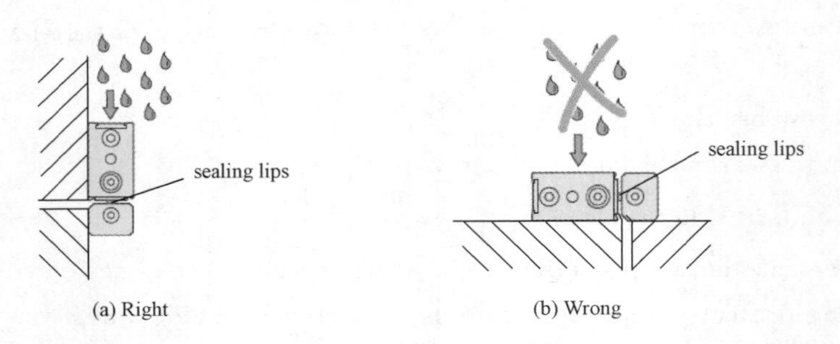

(a) Right (b) Wrong

Fig. 6-1-5 Right and Wrong Mounting Methods

Translation

模块 6 职场应用

任务 1 阅读产品说明书

1. CJX2 系列交流接触器

(1)适用范围

CJX2 系列交流接触器(图 6-1-1)是适用于交流 50 Hz 或 60 Hz,额定绝缘电压达690～1 000 V,在 AC-3 适用类别下额定工作电压为 380 V、额定工作电流达 95 A 的电力系统。它可供远距离接通和分断电路、频繁启动和控制交流电动机之用,也可与适当的热继电器或电子保护装置组合成电磁启动器,以防止电路或交流电动机发生过载。

图 6-1-1 CJX2 系列交流接触器

(2)结构特点

CJX2 系列交流接触器具有体积小、质量轻、寿命长和功耗小的特点。它可按要求组合派生,并可配置辅助触头组、空气延时头和机械联锁机构等附件,组成延时接触器、机械联锁接触器和星三角启动器。它还可以和热继电器直接插接，组成电磁启动器。

(3)主要技术参数(表 6-1-1)

表 6-1-1 主要技术参数

型号	额定工作电流/A	额定工作电压/V	额定绝缘电压/V	电寿命/10^4次	操作频率/h^{-1}	线圈功率/(V·A)		可接通负载
						吸合	保持	
CJX2-9	9	380	660	100	1 200	70	8	6 V 10 mA
CJX2-32	32	380	660	80	600	110	11	

2. 数控机床用直线光栅尺

(1)直线光栅尺的优点

直线光栅尺(图 6-1-2)测量直线轴的位置时不需要附加的机械转换元件。用直线光栅尺控制位置的控制环包括全部进给机构。安装在滑板上的直线光栅尺可以检测出机械转换误差并通过控制电路予以修正。它能消除许多潜在的误差源：

- 由滚珠丝杠的温度特性导致的位置误差；
- 滚珠丝杠反向间隙；
- 由滚珠丝杠的螺距误差导致的运动误差。

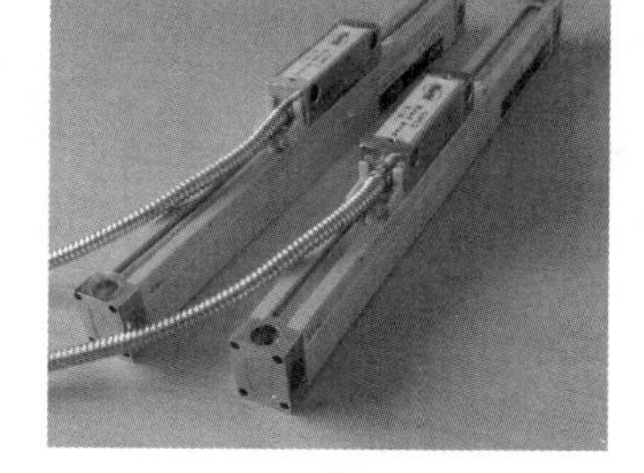

图 6-1-2 直线光栅尺

(2)机械结构

用于数控机床的直线光栅尺采用封闭式结构。铝制外壳保护直线光栅尺、扫描单元及其轨道免受切屑、灰尘和切削液的影响；弹性密封条密封外壳；扫描单元在摩擦力很小的轨道上运动，轨道内置在直线光栅尺上(图 6-1-3)。

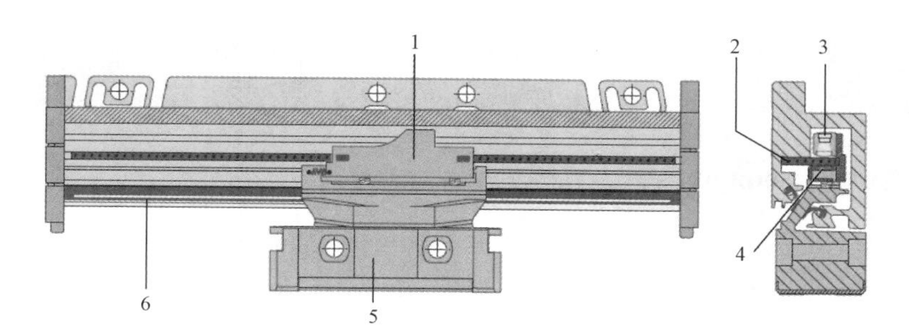

图 6-1-3 直线光栅尺的机械结构

1—扫描单元；2—直线光栅尺；3—光源；4—光电池；5—安装板；6—密封条

(3)安装指南

安装封闭式直线光栅尺非常简单，只需在多点位置将直线光栅尺与机床导轨对正，也可以用限位面或限位销对正直线光栅尺与机床导轨。安装辅助件时已将直线光栅尺和扫描单元的适当间隙以及横向公差调整正确。安装方法如图 6-1-4 所示。

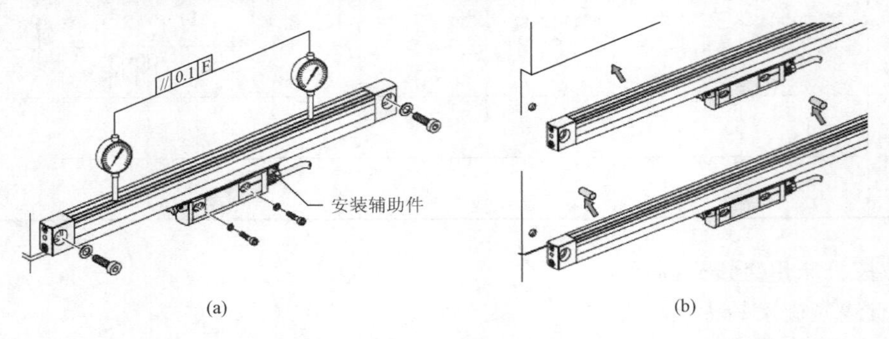

图 6-1-4　安装直线光栅尺

应将直线光栅尺的全长固定在被加工表面上，特别是有高动态性能要求时。安装时应将密封条朝下并远离溅水的方向(图 6-1-5)。

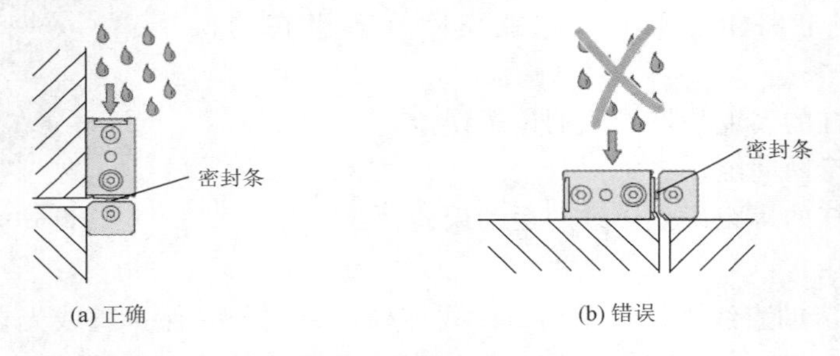

图 6-1-5　正确和错误的安装方法

Task 2　Applying for a Job

1. Job Advertisement

Star Light Company is a modern CNC machine manufacturing corporation. We have over 1 000 employees. We have successfully developed our position in the market in China. To further expand our market, we are seeking high-skilled people to join us in the following positions:

Engineering

Job description:

- CNC programming and program debugging;
- CNC equipment adjusting and maintaining;

- Technical service and support for users.

Job requirements:

- Bachelor's degree of electromechanical engineering, CNC background is preferred;
- Over 3 years' working experience in the CNC machining field;
- Familiar with CAD/CAM;
- Good ability of reading and speaking English.

Technician

Job description:

- Read and understand drawings, machining process, etc. ;
- Read and understand programs in FANUC system;
- Must be able to run more than one machine during cycle time.

Job requirements:

- Graduated from college majoring in electromechanical engineering;
- Minimum 2 years' experience in CNC operation;
- Good communicative skill and team work spirit;
- Age below 30.

2. Application Letter

An application letter should usually include the following:

(1)Job or position you want;

(2)Brie introduction of education and experience;

(3)Achievements;

(4)Skills;

(5)Anything that might help the reader decide if you are the kind of person they want.

Sample application letter:

Re: Application for the technician

Dear Sir,

Your advertisement for a technician on the April 10th *China Daily* interested me because the position that you described sounds exactly like the kind of job I am seeking for.

I have worked as a technician in GSK Company for the past 3 years and have gained rich experience in CNC operation and maintenance. Furthermore, I have recently completed an advanced level course about CNC programming and operation.

Currently, I am seeking for opportunities in a sizable and reputable organization like you to continue my career. I am confident that my practical experience and education background will show you that I can fulfill the requirements of your company.

Enclosed is my résumé. If I were favored with an interview, I would be grateful. Please contact me at 15236973697 any time. Thank you for your time and consideration!

Yours sincerely,
Zhang Hao

3. Personal Résumé

A résumé is the first step to present oneself to a potential employer. It opens the door to a job interview. A good résumé doesn't necessarily record all the information. The items in a résumé usually include personal details, education background and working experience, etc..

Sample personal résumé:

Personal Details

Name: Zhang Hao

Date of Birth: October 10th, 1997

Address: No. 499, Sanhao Rd., Hongqiao District, Tianjin, 300122

Telephone: 15236973697

E-mail: tjyjxy04@126.com

Marital Status: Single

Health: Excellent

Hobbies: Drawing, photography

Education Background

Sept., 2012～June, 2015: ××Middle School

Sept., 2015～June, 2018: Electromechanical Department of Tianjin Metallurgical Vocation & Technology Institute

Main Courses

Engineering Drawing	Mechanical Engineering
Analogue Electronics	Digital Electronics
Machining Process	Hydraulic and Pneumatic Drive
Electro-mechanical Control	CNC Systems
CNC Programming	CNC Machine Operation Practice
CAD/CAM	Enterprise Management

Fault Diagnosis and Maintenance for CNC Machines

Skills

PRETCO-(A)

CAD/CAM(advanced level)
CNC programming and operation(advanced level)
Familiar with office automation software
Clean driving license

Working Experiences

Summer, 2017: As a salesman of electromechanical products
2018～Present: Technician of CNC machine

Translation

任务2 求 职

1. 招聘广告

星光公司是一家现代数控机床制造企业。公司拥有1 000多名员工,已成功立足于中国市场。为了进一步拓展市场,我们正在寻觅高技术人才加盟,具体职位如下:

工程师

职位描述:

- 数控编程和程序调试;
- 数控设备调整和维护;
- 为用户提供技术服务和支持。

职位需求:

- 机电工程学士学位,有数控专业背景者优先;
- 在数控加工领域有3年以上的工作经验;
- 熟悉CAD/CAM;
- 良好的英语阅读、对话能力。

技术员

职位描述:

- 能读懂图纸、加工工艺等;
- 能读懂FANUC系统程序;
- 能同时操作多台机床。

职位需求：
- 机电工程专科毕业；
- 至少 2 年数控机床操作经验；
- 良好的沟通技巧和团队合作精神；
- 30 岁以下。

2. 求职信

求职信一般应包括如下内容：

(1)申请的职位；

(2)简单介绍教育和经历；

(3)成就；

(4)技能；

(5)有助于读者决定你是否是合适人选的其他内容。

求职信举例：

求职意向:技术员

尊敬的先生：

贵公司在 4 月 10 日《中国日报》上刊登的招聘技术员的广告吸引了我，因为广告上所描述的职位恰好是我一直在寻找的。

作为一名技术员，我在 GSK 公司已经工作了 3 年，在数控操作和维修方面积累了丰富的经验。此外，我最近已经学完了数控编程和操作的高级课程。

我一直渴望在像贵公司这样规模大且知名度较高的公司工作。我相信我的实践经验和教育背景将向贵公司表明我能够满足贵公司的要求。

随信附上我的简历。如蒙面试，我将不胜感激。可以随时拨打 15236973697 联系我。感谢您抽出时间并给予考虑！

此致

张浩

3. 个人简历

简历是向用人单位展示自己的第一步，它在求职面试中发挥着敲门砖的作用。一份好的简历不必面面俱到，它通常包括个人情况、教育背景和工作经历等内容。

个人简历举例：

◎个人情况

姓名：张浩

出生日期：1997 年 10 月 10 日

地址：天津市红桥区三号路499号，300122
电话：15236973697
电子邮箱：tjyjxy04@126.com
婚姻状况：未婚
健康状态：良好
业余爱好：绘画，摄影

◎教育背景

2012年9月至2015年6月：××中学
2015年9月至2018年6月：天津冶金职业技术学院机电系

◎主要课程

工程制图	机械工程
模拟电子	数字电子
机械加工工艺	液压与气压传动
机电控制	数控系统
数控编程	数控机床操作实习
计算机辅助设计与制造	企业管理
数控机床故障诊断与维修	

◎技能

高等学校英语应用能力考试A级
CAD/CAM（高级）
数控编程与操作（高级）
熟练使用办公自动化软件
有良好记录的驾驶执照

◎工作经历

2017年夏：机电产品销售员
2018年至今：数控机床技术员

Task 3 Exhibition Dialogue

1. Outside the Hall

Zhang: You are welcome to the CIMT. The opening ceremony is being performed now.

Smith: The scene is lively!

Smith: What's the CIMT?

Zhang： CIMT is short for China International Machine Tool exhibition.

Smith： What products will be shown on CIMT?

Zhang： The scope of CIMT includes the following： machines， tools， measurement equipments， automations for the industry and CAD/CAM， etc. .

Zhang： Look at those large balloons in the air with welcoming slogans on them.

Smith： This is very impressive indeed. It seems to be a big show.

Zhang： Exactly. The exhibitors include almost 2 200 companies from 97 countries. The registered visitors are already about 138 000 from all parts of the world and the number is expected to increase in two or three days.

Smith： Fantastic! Do you have such a big show every year?

Zhang： No. It has been held in China every 2 years ever since the first CIMT existed in 1989. The show has been recognized as one of the four most important machine exhibitions in the world.

Smith： I'm sure to come next time.

Zhang： You are welcome! Let me show you around.

Smith： It's very kind of you!

2. Inside the Hall

Zhang： This is the Exhibition Hall 2. It's composed of two sections. Here on display are some new homemade machines. Many of them have caught up with the technical levels of similar products made abroad.

Smith： Is this a big simultaneous five-axis CNC machine?

Zhang： Yes. This new machine has reached the advanced level in the world. It is suitable for aircraft industry.

Smith： I see.

Smith： Is this a vertical machining center?

Zhang： That's right. This is our recently developed product. The distinction of our product is economical and easy to operate. What's more， our service has been very well-received by our customers.

Smith： Good. That's what we want to hear. Could I see the specifications for the vertical machining center?

Zhang： Of course. Here is our product catalogue.

3. In the Negotiation Booth

Smith： We have studied your catalogue and have great interest in your vertical machining center. But your price has been found higher through repeated calculations.

Zhang： Considering the quality， I think our quotation is quite reasonable.

Smith： You know colleges don't have a lot of funds and this time we have the rare chance to get the appropriation from the Municipal Educational Commission and Financial Bureau to renovate our CNC training center. We have many types of equipment to buy and we are hard up for money. Please give us the most favorable price.

Zhang: How many do you want to order? And how would you make the payment?

Smith: We only need one of this type of machine, but we can buy other machines from your company in the future. We'll pay by installment.

Zhang: How about paying 30% upon signing the contract and then paying the rest after delivery of the goods?

Smith: According to our practice, we'll pay 60% after delivery and pay the remaining 10% after three months' trial use.

Zhang: We would like to have another new partner by reducing the unit price to $58 000. That's our bottom price. We will just use it as an advertisement.

Smith: It seems acceptable. When can you make the goods ready for shipment?

Zhang: Around mid-July.

Smith: Fine, we've decided to order it. When can we sign the contract?

Zhang: Tomorrow afternoon.

Smith: I hope it's a pleasant cooperation. See you tomorrow then.

Zhang: See you tomorrow.

Translation

任务3 展会对话

1. 展馆外

张： 欢迎来到CIMT。现在正在举行开幕式。

史密斯： 场面很热闹啊！

史密斯： 什么是CIMT？

张： CIMT是中国国际机床展览会的缩写。

史密斯： CIMT将展出哪些产品？

张： CIMT展出的产品包括机床、刀具、测量装置、工业自动化和CAD/CAM等。

张： 看那些在空中飘扬的大气球，上面写着欢迎标语。

史密斯： 真是让人印象深刻，看起来是一个大型的展会。

张： 一点也不错。来自97个国家的2 200个公司参展，已经登记的来自世界各地的参观者已达138 000人左右，在未来两三天内参观人数预计还会继续增加。

史密斯： 太棒了！每年都有这样大型的展会吗？

张： 不是，自从1989年的第一届中国国际机床展览会以来，每两年在中国举办一次。该展览会被公认为世界上最重要的四大机床展览会之一。

史密斯： 下次我肯定还会来的。

张： 欢迎，欢迎！我带你转转。

史密斯： 你真是太好了！

2. 展馆内

张：　这是 2 号展厅，由两部分组成。这里展出的是一些国产新机床，其中很多已经赶上国外同类产品的技术水平。

史密斯：　这是一台大型的 5 轴联动数控机床吗？

张：　是的。这台新机床已经达到了世界先进水平，适用于航空工业。

史密斯：　哦，是这样。

史密斯：　这是立式加工中心吗？

张：　是的。这是我们公司最新研制的产品，其特点是经济实用、操作简便。此外，顾客对我们的服务评价很高。

史密斯：　很好，这正是我们想要听到的。我能看一下立式加工中心的详细规格吗？

张：　当然可以，这是我们的产品目录。

3. 在谈判室

史密斯：　我们已经看过你们的样品，对你们的立式加工中心很感兴趣，但通过反复计算发现价格高了一点。

张：　考虑到质量，我认为我们的报价是很合理的。

史密斯：　您知道学校的经费不多，这次我们难得争取到市教委和财政局拨款来改造我们的数控实训中心的机会，要买的设备很多，经费紧张。希望贵方能给我们一个最优惠的价格。

张：　你们要买几台？怎样付款？

史密斯：　这种型号的机床我们只需要一台，但是今后我们可以从贵公司购买其他机床。我们将分期付款。

张：　签约时首付 30%，提货时再付清余款怎样？

史密斯：　按照我们的惯例，提货时付 60%，还有 10%的余款待试用三个月后付清。

张：　我们愿意把单价降到 5.8 万美元，这样我们可以成为新的合作伙伴。那是我们的底价，就算是做个广告吧。

史密斯：　似乎可以接受。贵方什么时候可以交货？

张：　7 月中旬。

史密斯：　好的，我们决定购买这台机床。我们什么时候可以签订合同？

张：　明天下午。

史密斯：　祝我们合作愉快。明天见。

张：　明天见。

学习启示

青涩年华化为多彩绽放，精益求精铸就青春信仰。我们要发挥自己的潜力，实现自己的价值。始终坚信，志高方能行远，有目标、敢奋斗，以豪情洒脱的气势，铸就无悔人生。

参考文献

1. 王兆奇，刘向红. 数控专业英语[M]. 2 版. 北京：机械工业出版社，2021.

2. 李玉萍. 机电英语[M]. 2 版. 北京：北京大学出版社，2021.

3 朱派龙. 图解机械工程英语[M]. 北京：化学大学出版社，2021.

4. 施平. 机械工程专业英语教程[M]. 5 版. 北京：电子工业出版社，2019.

5. 马庆芬，刘培启. 机械类专业英语应用教程[M]. 北京：机械工业出版社，2021.

6. 马玉录. 机械设计制造及其自动化专业英语[M]. 4 版. 北京：化学工业出版社，2020.

7. 沈延秀. 数控专业英语[M]. 北京：机械工业出版社，2021.

8. 陈青云，莫瑜. 数控专业英语[M]. 北京：机械工业出版社，2020.

9. 沈言锦. 机械专业英语[M]. 北京：机械工业出版社，2019.

10. 康兰，机械工程专业英语——交流与沟通[M]. 2 版. 北京：机械工业出版社，2019.

11. 于海祥，冯艳宏，张帆. 机电专业英语. [M]. 北京：中国铁道出版社，2017.

12. 韩林烨，关雄飞. 机械类专业英语[M]. 北京：机械工业出版社，2017.

13. 王鹏飞. 机电专业英语. [M]. 北京：北京理工大学出版社，2019.

14. 程福. 机械专业英语阅读教程[M]. 5 版. 大连：大连理工大学出版社，2018.

15. 董建国. 机械专业英语[M]. 3 版. 西安：西安电子科技大学出版社，2018.